T0254863

Lecture Notes in Computer Science 12874

Jing Chen · Minming Li ·
Guochuan Zhang (Eds.)

Frontiers of Algorithmics

International Joint Conference, IJTCS-FAW 2021
Beijing, China, August 16–19, 2021
Proceedings

Editors
Jing Chen
Stony Brook University
New York, NY, USA

Minming Li
Department of Computer Science
City University of Hong Kong
Kowloon, Hong Kong

Guochuan Zhang
College of Computer Science
Zhejiang University
Hangzhou, China

ISSN 0302-9743 ISSN 1611-3349 (electronic)
Lecture Notes in Computer Science
ISBN 978-3-030-97098-7 ISBN 978-3-030-97099-4 (eBook)
https://doi.org/10.1007/978-3-030-97099-4

LNCS Sublibrary: SL1 – Theoretical Computer Science and General Issues

This Springer imprint is published by the registered company Springer Nature Switzerland AG
The registered company address is: Gewerbestrasse 11, 6330 Cham, Switzerland

Preface

This volume contains the contributed, accepted papers presented at the 15th International Frontiers of Algorithmics Workshop (FAW 2021) and the Game Theory in Blockchain Track, these were part of the 2nd International Joint Conference on Theoretical Computer Science (IJTCS 2021) held during August 16–19, 2021, at Peking University in Beijing, China. Due to COVID-19, the conference was run in a hybrid mode.

The conference had both contributed talks submitted to the 15th International Frontiers of Algorithmics Workshop (FAW 2021) and the Blockchain track, and invited talks in focused tracks on Algorithmic Game Theory; Game Theory in Blockchain; Multi-agent Learning, Multi-agent System, Multi-agent Games; Learning Theory; Quantum Computing; Machine Learning and Formal Method; EconCS; and Conscious Turing Machine. Furthermore, a Female Forum, Young PhD Forum, Undergraduate Research Forum, and Young Faculty in TCS at CFCS session were also organized.

For the Frontiers of Algorithmics Workshop (FAW 2021), representing Track G of IJTCS, and the Blockchain track, representing Track B, the Program Committee, consisting of 23 top researchers from the field, reviewed nine submissions and decided to accept, respectively, 4 papers and 1 paper, and these are presented in this proceedings volume. [This volume does not include the invited talks presented in the other IJTCS tracks.] Each paper had at least three reviews, with additional reviews solicited as needed. The review process was conducted entirely electronically via EasyChair. We are grateful to EasyChair for allowing us to handle the submissions and the review process and to the Program Committee for their insightful reviews and discussions, which made our job easier.

Besides the regular talks, IJTCS-FAW had eight keynote talks by Manuel Blum and Lenore Blum (Carnegie Mellon University), Jing Chen (Stonybrook University), Wei Chen (Microsoft Research Asia), Xi Chen (Columbia University), Yijiia Chen (Shanghai Jiao Tong University), Holger Hermanns (Saarland University and Institute of Intelligent Software, Guangzhou), Shi Li (University at Buffalo), and Yitong Yin (Nanjing University).

In this front matter we show the organization of the FAW event, the overall IJTCS event, and also details for some of its tracks; we include the abstracts of the IJTCS keynote talks; and we include a list of the papers and talks presented in IJTCS Tracks A-F and H-I, and the Forums.

We are very grateful to all the people who made this meeting possible: the authors for submitting their papers, the Program Committee members and external reviewers for their excellent work, and all the keynote speakers and invited speakers. We also thank the Steering Committee for providing timely advice for running the conference. In particular, we would like to thank Peking University for providing organizational support. Finally, we would like to thank Springer for their encouragement and cooperation throughout the preparation of this conference.

August 2021

Jing Chen
Minming Li
Guochuan Zhang

Organization

Advisory Committee

Wen Gao (Chair)	Peking University
Hong Mei (Chair)	CCF President
Pingwen Zhang (Chair)	CSIAM President and Peking University
Yuxi Fu	Shanghai Jiao Tong University
Yunhao Liu	Tsinghua University
Xiaoming Sun	Chinese Academy of Sciences

Organizing Committee

Yunli Yang (Chair)	Peking University
Shuang Wu	Peking University
Hongyin Chen	Peking University
Yurong Chen	Peking University
Zhaohua Chen	Peking University
Zhijian Duan	Peking University
Wenhan Huang	Shanghai Jiao Tong University
Jichen Li	Peking University
Mengqian Zhang	Shanghai Jiao Tong University

Conference Chair

John E. Hopcroft	Cornell University

General Chair

Xiaotie Deng	Peking University

Program Committee Chairs

Xiaotie Deng	Peking University
Minming Li	City University of Hong Kong
Jian Li	Tsinghua University
Pinyan Lu	Shanghai University of Finance and Economics
Xiaoming Sun	Chinese Academy of Sciences

Track Chairs

Yukun Cheng	Suzhou University of Science and Technology (Chair of Track A: Algorithmic Game Theory)
Zhengyang Liu	Beijing Institute of Technology (Chair of Track A: Algorithmic Game Theory)
Zhihao Tang	Shanghai University of Finance and Economics (Chair of Track A: Algorithmic Game Theory)
Jing Chen	Stony Brook University (Chair of Track B: Game Theory in Blockchain)
Xiaotie Deng	Peking University (Chair of Track B: Game Theory in Blockchain)
Wenxin Li	Peking University (Chair of Track C: Multi-agent Learning, Multi-agent System, Multi-agent Games)
Haifeng Zhang	Chinese Academy of Sciences (Chair of Track C: Multi-agent Learning, Multi-agent System, Multi-agent Games)
Jian Li	Tsinghua University (Chair of Track D: Learning Theory)
Xiaoming Sun	Chinese Academy of Sciences (Chair of Track E: Quantum Computing)
Jialin Zhang	Chinese Academy of Sciences (Chair of Track E: Quantum Computing and Female Forum)
Lijun Zhang	Chinese Academy of Sciences (Chair of Track F: Machine Learning and Formal Method)
Minming Li	City University of Hong Kong (Chair of Track G: 15th Frontiers of Algorithmics Workshop)
Guochuan Zhang	Zhejiang University (Chair of Track G: 15th Frontiers of Algorithmics Workshop)
Biaoshuai Tao	Shanghai Jiao Tong University (Chair of Track H: EconCS)
Manuel Blum	Carnegie Mellon University, Emeritus (Chair of Track I: Conscious Turing Machine)
Lenore Blum	Carnegie Mellon University, Emeritus (Chair of Track I: Conscious Turing Machine)
Yurong Chen	Peking University (Chair of Track I: Conscious Turing Machine)
Zhaohua Chen	Peking University (Chair of Undergraduate Research Forum)
Yuqing Kong	Peking University (Chair of Female Forum and Young Faculty in TCS at CFCS)
Xiang Yan	Huawei (Chair of Young PhD Forum)
John E. Hopcroft	Cornell University (Chair of Young Faculty in TCS at CFCS)

Program Committee of Track A: Algorithmic Game Theory

Xiang Yan	Huawei, China
Zhengyang Liu	Beijing Institute of Technology
Xujin Chen	Chinese Academy of Sciences
Zhigang Cao	Beijing Jiaotong University
Jie Zhang	University of Southampton
Tao Xiao	Huawei, Germany
Yukun Cheng	Suzhou University of Science and Technology, China
Mingyu Guo	University of Adelaide
Weiran Shen	Renmin University of China
Ye Du	Southwestern University of Finance and Economics
Haoran Yu	Beijing Institute of Technology
Jinyan Liu	Beijing Institute of Technology
Zhihao Tang	Shanghai University of Finance and Economics
Xiaohui Bei	Nanyang Technological University
Zihe Wang	Renmin University of China
Qi Qi	Hong Kong University of Science and Technology

Program Committee of Track G: 15th Frontiers of Algorithmics Workshop

Vincent Chau	Southeast University
Lin Chen	Texas Tech University
Xujin Chen	Chinese Academy of Sciences
Qilong Feng	Central South University
Hiroshi Fujiwara	Shinshu University
Bundit Laekhanukit	Shanghai University of Finance and Economics
Guojun Li	Shandong University
Chung-Shou Liao	National Tsinghua University
Guohui Lin	University of Alberta
Aris Pagourtzis	National Technical University of Athens
Haitao Wang	Utah State University
Xiaowei Wu	University of Macau
Mingyu Xiao	University of Electronic Science and Technology of China
Yuhao Zhang	Shanghai Jiaotong University

Program Committee of Track B: Game Theory in Blockchain

Artem Barger	IBM, Israel
Hubert Chan	The University of Hong Kong
Bo Li	Hong Kong Polytechnic University
Georgios Piliouras	Singapore University of Technology and Design

Additional Reviewer

Chao Xu	University of Electronic Science and Technology of China, China

Keynote Speeches

Insights from the Conscious Turing Machine (CTM)

Manuel Blum and Lenore Blum

Carnegie Mellon University

Abstract. The quest to understand consciousness, once the purview of philosophers and theologians, is now actively pursued by scientists of many stripes. In this talk, we discuss consciousness from the perspective of theoretical computer science (TCS), a branch of mathematics concerned with understanding the underlying principles of computation and complexity, especially the implications of resource limitations. In the manner of TCS, we formalize the Global Workspace Theory (GWT) originated by cognitive neuroscientist Bernard Baars and further developed by him, Stanislas Dehaene, and others. Our principal contribution lies in the precise formal definition of a Conscious Turing Machine (CTM). We define the CTM in the spirit of Alan Turing's simple yet powerful definition of a computer, the Turing Machine (TM). We are not looking for a complex model of the brain nor of cognition but for a simple model of (the admittedly complex concept of) consciousness.

After defining CTM, we give a formal definition of consciousness in CTM. We then suggest why the CTM has the feeling of consciousness.

The perspective given here provides a simple formal framework to employ tools from computational complexity theory and machine learning to further the understanding of consciousness.

This is joint work of Manuel, Lenore and Avrim Blum.

Speculative Smart Contracts

Jing Chen

Stony Brook University and Algorand Inc.

Abstract. In this talk, I'll introduce Algorand's layer-2 speculative smart contract architecture and discuss the design principles behind it.

Optimization from Structured Samples—An Effective Approach for Data-Driven Optimization

Wei Chen

Microsoft Research Asia

Abstract. Traditionally machine learning and optimization are two different branches in computer science. They need to accomplish two different types of tasks, and they are studied by two different sets of domain experts. Machine learning is the task of extracting a model from the data, while optimization is to find the optimal solutions from the learned model. In the current era of big data and AI, however, such separation may hurt the end-to-end performance from data to optimization in unexpected ways. In this talk, I will introduce the paradigm of data-driven optimization that tightly integrates data sampling, machine learning, and optimization tasks. I will mainly explain two approaches in this paradigm, one is optimization from structured samples, which carefully utilizes the structural information from the sample data to adjust the learning and optimization algorithms; the other is combinatorial online learning, which adds feedback loop from the optimization result to data sampling and learning to improve the sample efficiency and optimization efficacy. I will illustrate these two approaches through my recent research studies in these areas.

Recent Developments in Property Testing of Boolean Functions

Xi Chen

Columbia University

Abstract. Over the past few decades, property testing has emerged as an important line of research in sublinear computation. At a high level, testers are ultra-fast randomized algorithms which determine whether a "massive object" satisfies a particular property while only inspecting a tiny portion of the object. In this talk, we will review some of the recent developments in property testing of Boolean functions.

AC^0 Circuits, First-Order Logic, and Well-Structured Graphs

Yijia Chen

Fudan University

Abstract. In this talk, I will explain some recent results on evaluating queries on well-structured graphs using AC^0 circuits (i.e., circuits of constant depth and polynomial size). We study those queries definable in monadic second-order logic (MSO), including NP-hard problems like 3-colorability and SAT. We exploit the well-known connection between AC^0 and first-order logic (FO) to design our circuits by developing some machinery to describe those well-structured graphs in FO. These results yield fast parallel algorithms on certain dense graphs.

Model-Based Digital Engineering and Verification of Intelligent Systems

Holger Hermanns

Saarland University

Abstract. From autonomous vehicles to Industry 4.0, increasingly computer programs participate in actions and decisions that affect humans. However, our understanding of how these applications interact and what is the cause of a specific automated decision is lagging far behind.

In this talk, we will introduce the life cycle of systems engineering: Design-Time, Run-Time, and Inspection-Time. And the RAMS (Reliability, Availability, Maintainability, and Safety) properties that we want to verify during the entire system life cycle. Then we will show how we reduce complexity by making reasonable assumptions and propose the Racetrack Verification Challenge. We will talk about how we apply deep statistical model checking, abstract interpretation, and other techniques to solve this challenge in our recent work.

Tight Online Algorithms for Unrelated Machine Load Balancing with Predictions

Shi Li

University at Buffalo

Abstract. The success of modern machine learning techniques has led to a surge of interest in using machine learned predictions to design better online algorithms for combinatorial optimization problems. This is called learning augmented online algorithms. One seeks to design a prediction about the online instance and an online algorithm which is given the prediction upfront. They should satisfy: a) usefulness, which means the prediction allows the online algorithm to achieve a better competitive ratio, b) robustness, which means the performance of the algorithm deteriorates smoothly as the noise of prediction increases, and c) learnability, which means the predicted information can be learned efficiently from past instances.

In this talk, I will present our results for the online load balancing problem in this model. That is, we allow some prediction about the instance to be given to the online algorithm. First, we design deterministic and randomized online rounding algorithms for the problem in the unrelated machine setting, with $O(\log m/\log \log m)$- and $O(\log \log m/\log \log \log m)$-competitive ratios. They respectively improve upon the previous ratios of $O(\log m)$ and $O(\log^3 \log m)$ of Lattanzi et al., and match their lower bounds. Second, we extend their prediction scheme from the restricted assignment setting to the unrelated machine setting. Finally, we consider the learning model introduced by Lavastida et al., and show that under the model, the prediction can be learned efficiently with a few samples of instances.

The talk is based on my joint work with Jiayi Xian.

Fast Sampling Constraint Satisfaction Solutions via the Lovász Local Lemma

Yitong Yin

Nanjing University

Abstract. This talk will cover some recent advances in sampling solutions of constraint satisfaction problems (CSPs) through the Lovász local lemma. We will talk about Markov chain based algorithms for sampling almost uniform CSP solutions, inspired by an LP based approach for approximately counting the number of CSP solutions due to Ankur Moitra. Assuming "local lemma"-like conditions, these Markov chain based sampling algorithms are accurate and can run in a nearly linear time in the number of variables for a broad class of CSPs.

Contents

Papers and Talks Presented in IJTCS Tracks A-F and H-I, and the Forums

Invited Talks, Track A (Algorithmic Game Theory)

Strongly Robust Exotic Option Pricing Via No-Regret Learning [Ye Du]
Learning Utilities and Equilibria in Non-Truthful Auctions [Hu Fu]
Assortment games under Markov Chain Choice Model [Changjun Wang]
Abstract Market Games with Gross Substitutes/Complements [Zhigang Cao]

Invited Talks, Track B (Game Theory in Blockchain)

Framework and algorithms for identifying honest blocks in blockchain [Xu Wang]
Evolutionary Equilibrium Analysis for Decision on Block Size in Blockchain Systems [Jinmian Chen, Yukun Cheng, and Zhiqi Xu]
Enabling an alternative trust model with Blockchain DB [Artem Barger]
Sublinear-Round Byzantine Agreement under Corrupt Majority [Hubert Chan]
Maximal Information Propagation via Lotteries [Bo Li]
Dynamical Analysis of the EIP-1559 Ethereum Fee Market [Georgios Piliouras]
Block-CES: Multi-institution collaborative credit evaluation system based on blockchain [Yuncheng Qiao]
An intelligent electric vehicle charging system for new energy companies based on consortium blockchain [Zhengtang Fu]

Invited Talks, Track C (Multi-agent Learning, Multi-agent System, Multi-agent Games)

Integrating knowledge-based and data-driven paradigms for collective intelligence decision making: algorithms and experiments [Zhiqiang Pu]
MALib: A Parallel Framework for Population-based Multi-agent Reinforcement Learning [Ying Wen]
Reinforcement Learning for Incomplete Information Games [Chao Yu]
Dealing with Non-transitivity in Two-player Zero-sum Games [Yaodong Yang]

Invited Talks, Track D (Learning Theory)

Provable Representation Learning [Simon S. Du]
Settling the Horizon-Dependence of Sample Complexity in Reinforcement Learning [Lin Yang]
The Power of Exploiter: Provable Multi-Agent RL in Large State Spaces [Chi Jin]

Invited Talks, Track E (Quantum Computing)

Quantum Artificial Intelligence—The Recent Advances [Dongling Deng]
Perturbative quantum simulation [Xiao Yuan]

Quantum Algorithms for Semidefinite Programs with Applications to Machine Learning [Tongyang Li]

Invited Talks, Track F (Machine Learning and Formal Method)

Explaining Deep Neural Networks using Statistical Fault Localization [Youcheng Sun]
Safety Verification of Neural-Network Controlled Systems [Chao Huang]
Robustness Analysis for DNNs from the Perspective of Model Learning [Cheng-chao Huang]
BDD4BNN: A BDD-based Quantitative Analysis Framework for Binarized Neural Networks [Fu Song]

Invited Talks, Track H (EconCS)

Candidate Selections with Proportional Fairness Constraints [Shengxin Liu]
Multiplayer bandit learning, from competition to cooperation [Simina Brânzei]
Fair division of indivisible goods [Jugal Garg]
Eliciting Expert Information without Verification [Grant Schoenebeck]

Invited Talks, Track I (Conscious Turing Machine)

Mathematical Foundations of CTM [Rui Ding and Lingxiao Zheng]
Doorway CTM: Explaning forgetting via Conscious Turing Machine [Lan Lyu, Chang Ma, and Bowen Su]
MusiCTM: On Music Composing by CTM [Shuchen Li]

Undergraduate Research Forum

The Smoothed Complexity of Computing Kemeny and Slater Rankings [Weiqiang Zheng]
Optimal Multi-Dimensional Mechanisms are not Local [Zixin Zhou]
Timely Information from Prediction Markets [Chenkai Yu]
Random Order Vertex Arrival Contention Resolution Schemes For Matching, With Applications [Jinzhao Wu]
Maximizing Surprise with a Bonus [Zhihuan Huang]
Recent Progress on Bandits and RL with Nonlinear Function Approximation [Jiaqi Yang]
Inverse-Exponential Correlation Bounds and Extremely Rigid Matrices from a New Derandomized XOR Lemma [Xin Lyu]
Faster algorithms for distance sensitivity oracles [Hanlin Ren]

Female Forum

Optimal Advertising for Information Products [Shuran Zheng]
Debiasing Evaluations That Are Biased by Evaluations [Jingyan Wang]
Designing a Combinatorial Financial Options Market [Xintong Wang]

Young PhD Forum

Making Method of Moments Great Again? – How can GANs learn distributions [Zehao Dou]
Cursed yet Satisfied Agents [Juntao Wang]
Convergence Analysis of No-Regret Bidding Algorithms in Repeated Auctions [Zhe Feng]

Counting solutions to random CNF formulas [Kuan Yang]
Optimal Mixing of Glauber Dynamics via Spectral Independence Approach [Zongchen Chen]
Libra: Succinct Zero-Knowledge Proofs with Optimal Prover Computation [Jiaheng Zhang]
Proving Expected Sensitivity of Probabilistic Programs with Randomized Variable-Dependent Termination Time [Peixin Wang]
Maximin Fairness with Mixed Divisible and Indivisible Goods [Xinhang Lu]
Almost Tight Approximation Hardness for *k*-DST [Chao Liao]

Young Faculty in TCS at CFCS

Streaming Algorithms for Geometric Steiner Forest [Shaofeng Jiang]
Efficient Document Exchange and Error Correcting Codes with Asymmetric Information [Kuan Cheng]
Optimizing Multi-task Peer Prediction [Yuqing Kong]

Full Papers

Pool Block Withholding Attack with Rational Miners

Ni Yang[1], Chenhao Wang[2,3](✉), and Bo Li[4]

[1] Department of Computer Science, City University of Hong Kong, HKSAR, China
niyang3-c@my.cityu.edu.hk

[2] Advanced Institute of Natural Sciences, Beijing Normal University, Zhuhai, China
wangch@amss.ac.cn

[3] BNU-HKBU United International College, Zhuhai, China

[4] Department of Computing, The Hong Kong Polytechnic University, HKSAR, China
comp-bo.li@polyu.edu.hk

Abstract. We revisit pool block withholding (PBW) attack in this work, while taking miners' rationality into consideration. It has been shown that a malicious mining pool in Bitcoin may attack other pools by sending some dummy miners who do not reveal true solutions but obtain reward portions from attacked pools. This result assumes that the infiltrating miners are always loyal and behave as the malicious pool desires; but why? In this work, we study rational infiltrating miners who behave by maximizing their own utilities. We characterize the infiltrating miners' optimal strategies for two kinds of betraying behaviors depending on whether they return to the original pool or not. Our characterizations show that the infiltrating miners' optimal strategy is simple and binary: they either betray or stay loyal all together, which depends on the total amount of them sent out by the malicious pool. Accordingly, we further compute the pool's optimal attacking strategy given the miners' strategic behaviors. Our experiments also show that by launching PBW attacks, the benefit to a pool is actually incremental.

1 Introduction

With the popularity of cryptocurrencies such as Bitcoin [15], the recent decade has seen remarkable theoretical achievements and industrial technologies about blockchain [5,14,21,23]. At the heart of blockchain, there is a tamperproof distributed ledger storing all the transactions that all participants in a Peer-to-Peer network reach consensus.

Transactions are stored in blocks, and anyone can join the system to verify the transactions' legitimacy and pack them into blocks. Due to the openness of blockchain, Bitcoin uses a proof-of-work mechanism: it requires participants to solve a cryptographic puzzle, which needs significant amount of computational

This work is supported by The Hong Kong Polytechnic University (Grant No. P0034420).

J. Chen et al. (Eds.): IJTCS-FAW 2021, LNCS 12874, pp. 3–22, 2022.
https://doi.org/10.1007/978-3-030-97099-4_1

resources. The participant who successfully finds the solution can attach a new block to the chain and get rewarded with Bitcoins.

The process of generating a block and solving a problem is called *mining*, and the participants are *miners*. Informally, mining consists in repeatedly computing hashes of variants of a data structure, until one is found whose numerical value is low enough (i.e., solution, also called *full proof-of-work*). The process of calculating hashes individually and receiving reward entirely when finding a valid block is known as *solo mining*. Roughly, the probability that a miner gets the full proof of work is proportional to her computational power over the whole power in the system, which means that the variance of finding it is substantially high. A natural process leads miners to form *pools*, where they can aggregate their power and share the rewards whenever one of them creates a block.

Each pool distributes the pool reward to all its members proportionally to their computational power. In order to estimate their true computational power, the pool manager may use *partial proof-of-work*, which is an approximate but not precise solution. That is, each member's final reward is proportional to the number of partial proof-of-work submitted by her over that number by all. There are more mining pool reward schemes which can be found in [17]. No matter which scheme is adopted, in expectation or in the long run, the reward to a miner does not change by joining a pool; however, she obtains more stable income.

Although Bitcoin has already been a remarkable feat, it remains an active research area on many fronts. For example, Rosenfeld [17] initiated the discussion of *block withholding attack*, where a pool member can sabotage an open pool by delaying the submission of proof-of-works. Later, Eyal [8] further proved that a malicious mining pool may have an incentive to send some of its miners to infiltrate the other pools, which is called *pool block withholding (PBW) attack*. These infiltrating miners only submit their partial proof-of-works to the infiltrated pools and discard all full proof-of-works. That is, they pretend to work, but make no contribution to these pools. Seemingly, the malicious pool does not benefit from this attack as it sacrifices some of its computational power. However, as the infiltrated pools are not able to recognize infiltrating miners from their own honest ones, the infiltrating miners can still obtain their reward portions from these pools. It is shown in [8] that a pool is always beneficial to launch PBW attacks if the infiltrating miners bring their stealing rewards back.

However, to the best of our knowledge, we have not observed any PBW attack in practice. Actually, the above result crucially depends on the assumption that all infiltrating miners are loyal to the original pool. Whether this assumption is true or not? To figure out this problem, in this work, we revisit the effects of PBW attacks by taking the rationality of infiltrating miners into consideration. Though some works, e.g. [11,22], consider the phenomenon of betrayal, they assume a fixed fraction of infiltrating miners betray the original pool. In our model, these infiltrating miners sent by the pool are not compulsorily loyal or betrayal to the original pool, by which we mean they arbitrarily decide whether or not to submit the full proof-of-works to maximize their own rewards. Given the rational behavior of miners, we reanalyze the malicious pool's optimal PBW attack strategy.

1.1 Main Results

In this work, we consider the rational behavior of infiltrating miners in a strong sense, where they collectively decide that an arbitrary fraction of them betray the malicious pool, and the remaining of them stay loyal. The goal of these infiltrating miners is to maximize their total reward. The only assumption we make is that submitting full proof-of-work or not does not affect their expected reward portion, which can be justified by the fact that the number of full proof-of-works is negligible compared with that of partial proof-of-works. We further distinguish two types of betraying behaviors, both of which submit full proof-of-works to the attacked pool, but one brings the stealing reward back to the malicious pool and the other does not.

Betray-and-Return (BR) Model. A bit more formally, in a BR model, the infiltrating miners are split into two groups, one of which actually submit full proof-of-works, and the other does not. After the attack, both of them return to the malicious pool and pretend that they have not submitted full proof-of-works. This model also captures the scenario when the malicious pool does not or is not able to monitor all infiltrating miners.

For the BR model, we show that the infiltrating miners' optimal strategy is simple: there exists a threshold such that when the malicious pool's attacking power is above this threshold, then the infiltrating miners' best response is to completely betray; otherwise, they will be completely loyal. This result can be extended to the case of attacking multiple pools simultaneously, based on which, we are able to characterize the optimal attacking strategies for malicious pools. We also conduct extensive experiments to illustrate how much more revenue a malicious pool can obtain by attacking others when miners are rational. Our experiments show that, no matter how large the malicious pool is (regarding it computation power compared with others), its revenue is increased by no more than 10%. These results are presented in Sect. 3.

Betray-and-Not-Return (BNR) Model. In the BNR model, although infiltrating miners are also split into two groups, one group directly leaves the malicious pool by submitting full proof-of-works and not returning, and only the other loyal group return to the malicious pool. Accordingly, the BNR model captures the scenario when the malicious pool can detect whether an infiltrating miner betrayed or not.

For the BNR model, when the malicious pool is only attacking one of the other pools, we prove exactly the same result with BR model. When it attacks more than 1 pools, the problem becomes more complicated. As shown by our experiments, there still exist thresholds which are the turning point for infiltrating miners to be completely betrayer or completely loyal. However, these thresholds are much smaller than the ones in BR models. That is, the infiltrating miners get more easily to betray the malicious pool, and thus the malicious tends to not attack other pools.

We remark that, in this paper, we mainly build the BR and BNR model and analyze the optimal strategy of a single pool given others' actions instead of simultaneous actions of multiple pools, which is far more complicated.

1.2 Related Work

Block Withholding Attack. Rosenfeld [17] first introduced the block withholding attack, which is later generalized to a formal concept of the sabotage attack [7]. Courtois and Bahack [7] also proposed a method which can tolerate a substantial percentage of miners engaged in it without being detected by small pools or individual miners. Eyal [8] then took a game theoretic perspective and considered mutual attacks among two pools. He proved that honest mining (i.e., no attack) is always not a Nash equilibrium and mutual attacks harm both of them, which is thus called the miner's dilemma. Following [8], Alkalay-Houlihanand and Shah [1] analyzed the pure price of anarchy (PPoA) of the game which bounds the loss of computational resources. Similarly, Luu et al. [13] also considered the case of attacking multiple pools and provided an algorithmic strategy by defining the power splitting game. In order to get closer to the real-world scenario, Ke et al. [10] presented a detailed analysis of the dynamic reward of the block-withholding attacker while considering the difficulty changing and proposed a novel adversarial attack strategy named the intermittent block withholding attack. Compared to previous works, Chen and Wang [22] introduced an extended model of 2-player miner's dilemma with betrayal assumption. They proved the existence and uniqueness of pure Nash equilibrium of this game and gave a tight bound of $(1, 2]$ for PPoA in a more general assumption.

Preventing Block Withholding Attack. Since block withholding attack could be harmful to open pools, effort has been devoted to revising the Bitcoin protocol or pool reward scheme to discourage or avoid such attacks. For example, Schrijvers et al. [20] designed a new incentive compatible reward function and showed proportional mining rewards are not incentive compatible. A special reward scheme is proposed in [3], which discourages the attackers by granting additional incentive to a miner who actually finds a block. An attacker who never submits a valid block to the pool will never receive the special reward and her revenue will be less than her expectation. Later, Bag et al. [2] analyzed the reward function of rogue miners and pools. They proposed that the pool manager are able to keep part of the revenue of each miner until they submit a valid proof-of-work. Recently, Chen et al. [6] revised the reward scheme by allowing pool managers to directly deduct part of their total mining rewards before the reward distribution. They showed that for properly designed deductions, honest mining is always a Nash equilibrium.

Other Security Attacks. Bitcoin has also been shown to be vulnerable for other attacks besides block withholding. To name a few of them, Rosenfeld [18] and Pinzón and Rocha [16] proposed and analyzed the double-spending attack; Eyal and Sirer [9] and Sapirshtein et al. [19] proposed and generalized the selfish mining attack; Bonneau [4] introduced bribery attack, and Liao and Katz [12] carried out a corresponding game-theoretic analysis.

2 Models and Preliminaries

We first formally define our problems in this section. Though we will mainly focus on the strategic behavior of one particular mining pool, we introduce a general model where every mining pool is abstracted as an agent, for ease of exposition. As we consider the static and one-shot decision, the mining power of each pool and the total mining power in the system are assumed to be fixed.

Suppose there are in total n mining pools in the system, and the set of all pools is denoted by $N = \{1, \dots, n\}$. Each pool $i \in N$ has mining power $m_i \in \mathbb{R}^+$. The mining power controlled by solo miners (who are not in any pool) is m_0. Let $m = m_0 + m_1 + \cdots + m_n$ be the total mining power in the system. We focus on one particular pool, say pool 1, and analyze her strategic behavior and utility, given that the miners are rational, while other pools are assumed to be nonstrategic.

A PBW attacking strategy of pool 1 is defined as a vector $\mathbf{x} = (x_i)_{i=2,\cdots,n}$ with $x_i \geq 0$ and $\sum_{i=2}^{n} x_i \leq m_1$, that is, pool 1 attacks pool i by sending infiltrating miners with exactly x_i mining power, who are supposed to withhold the true solutions in the mining activity. However, rational (infiltrating) miners will not simply follow pool 1's instructions and may betray if it brings more benefit to themselves. In this work, we model two kinds of betraying behaviours: *betray-and-return* (BR), where betraying miners bring their reward share back to pool 1, and *betray-and-not-return* (BNR), where betraying miners do not.

2.1 Betray-and-Return Model

In a BR model, the betraying miners submit true solutions to their infiltrated pools, but bring their reward share back to pool 1 and pretend they have not. Accordingly, pool 1 is not able to verify if the infiltrating miners have betrayed or not.

Given attacking strategy $\mathbf{x}$ of pool 1, we assume that each x_i mining power (which attacks pool i) is controlled by a single infiltrating miner i, which can be well justified as all miners controlling x_i can form a collision and determine together an overall reaction. The reaction profile of all infiltrating miners is denoted by $\boldsymbol{\alpha} = (\alpha_i)_{i=2,\cdots,n}$, where $0 \leq \alpha_i \leq x_i$. Here α_i is the mining power loyal to pool 1, which does not contribute to pool i, and $x_i - \alpha_i$ is the mining power betraying pool 1, which contributes to pool i.

Let R_i be the *direct mining reward* of pool i from the system, which is the ratio of the effective mining power contributed by pool i, over the total effective power. Formally,

$$R_1(\mathbf{x}, \boldsymbol{\alpha}) = \frac{m_1 - \sum_i x_i}{m - \sum_i \alpha_i}, \tag{1}$$

and for $i = 2, \cdots, n$,

$$R_i(\mathbf{x}, \boldsymbol{\alpha}) = \frac{m_i + x_i - \alpha_i}{m - \sum_j \alpha_j}. \tag{2}$$

When $\mathbf{x}$ and $\boldsymbol{\alpha}$ are clear in the context, we write R_1 and R_i for short.

Besides the direct mining reward, pool 1 also have *infiltrating reward* $x_i \cdot r_i^p$ from each pool $i \geq 2$, where

$$r_i^p(\mathbf{x}, \boldsymbol{\alpha}) = \frac{R_i(\mathbf{x}, \boldsymbol{\alpha})}{m_i + x_i}$$

is the reward per unit mining power paid by pool i. Thus, pool 1's total reward (utility) is denoted by

$$u_1(\mathbf{x}, \boldsymbol{\alpha}) = [m_1 - \sum_{i=2}^{n}(x_i - \alpha_i)] \cdot r_1^p(\mathbf{x}, \boldsymbol{\alpha}),$$

where betraying miners' utilities are excluded, and

$$r_1^p(\mathbf{x}, \boldsymbol{\alpha}) = \frac{R_1(\mathbf{x}, \boldsymbol{\alpha}) + \sum_{i=2}^{n} x_i \cdot r_i^p(\mathbf{x}, \boldsymbol{\alpha})}{m_1}$$

is the unit reward of pool 1.

Finally, the utility of each infiltrating miner $i \geq 2$ is

$$u_i(\mathbf{x}, \boldsymbol{\alpha}) = x_i \cdot r_1^p(\mathbf{x}, \boldsymbol{\alpha}),$$

that is, x_i times the unit reward received from pool 1. Since x_i is regarded as a constant by miner i, it implies that i's utility is maximized when $r_1^p(\mathbf{x}, \boldsymbol{\alpha})$ is maximized.

2.2 Betray-and-Not-Return Model

In a BNR model, the betraying miners completely deviate from pool 1 by submitting true solutions to their infiltrated pools, and not bringing their reward share back to pool 1. Thus, pool 1 knows how many of the infiltrating powers have betrayed.

Similarly, given attacking strategy $\mathbf{x}$ of pool 1, we use $\boldsymbol{\alpha} = (\alpha_i)_{i=2,\cdots,n}$ with $0 \leq \alpha_i \leq x_i$, to denote infiltrating miners' reaction profile. The direct mining rewards are still R_1 and R_i as in (1) and (2), but pool 1's utility is different. Since betraying miners do not bring their reward share back, pool 1's infiltrating reward is only $\alpha_i \cdot r_i^d$ from each pool $i \geq 2$, where

$$r_i^d(\mathbf{x}, \boldsymbol{\alpha}) = r_i^p(\mathbf{x}, \boldsymbol{\alpha}) = \frac{R_i(\mathbf{x}, \boldsymbol{\alpha})}{m_i + x_i}.$$

Because pool 1 can identify the infiltrating powers that have betrayed, the utility of pool 1 is

$$u_1(\mathbf{x}, \boldsymbol{\alpha}) = R_1 + \sum_{i=2}^{n} \alpha_i \cdot r_i^d(\mathbf{x}, \boldsymbol{\alpha}) = R_1 + \sum_{i=2}^{n} \frac{\alpha_i}{m_i + x_i} R_i,$$

and the unit reward of pool 1 is

$$r_1^d(\mathbf{x}, \boldsymbol{\alpha}) = \frac{R_1 + \sum_{i=2}^n \alpha_i \cdot r_i^d(\mathbf{x}, \boldsymbol{\alpha})}{m_1 - \sum_{j=2}^n (x_j - \alpha_j)}.$$

Accordingly, the utility of each infiltrating miner i consists of two parts: one is the mining reward in pool i, i.e., $(x_i - \alpha_i) \cdot r_i^d$, and the other is the loyal share from pool 1, i.e., $\alpha_i \cdot r_1^d$. Thus the utility is

$$u_i(\mathbf{x}, \boldsymbol{\alpha}) = (x_i - \alpha_i) \cdot r_i^d(\mathbf{x}, \boldsymbol{\alpha}) + \alpha_i \cdot r_1^d(\mathbf{x}, \boldsymbol{\alpha}).$$

In both BR and BNR models, we are interested in the *best reaction* of infiltrating miner i: $\alpha_i^* \in \arg\max_{\alpha_i \in [0, x_i]} u_i(\mathbf{x}, \boldsymbol{\alpha})$, given pool 1's strategy $\mathbf{x}$ and all other miner j's ($j \neq i$) reaction α_j. Moreover, we are interested in pool 1' optimal strategy $\mathbf{x}^* := \arg\max_{\mathbf{x}} u_1(\mathbf{x}, \boldsymbol{\alpha}_{\mathbf{x}})$, where $\boldsymbol{\alpha}_{\mathbf{x}}$ is the best reaction profile of infiltrating miners w.r.t. $\mathbf{x}$.

3 Betray-and-Return Model

In this section, we consider the Betray-and-Return model. We first study the strategic behavior of a single infiltrating miner i, and then study pool 1's optimal strategy for attacking.

3.1 A Single Miner's Best Reaction

We arbitrarily fix x_i for $i \geq 2$ such that $\sum_{i=2}^n x_i \leq m_1$, and fix $\alpha_i \leq x_i$ for $i \geq 3$. We study the best reaction α_2^* for infiltrating miner 2, that is,

$$\alpha_2^* \in \arg \max_{\alpha_2 \in [0, x_2]} u_2(\mathbf{x}, \boldsymbol{\alpha}) = \arg \max_{\alpha_2 \in [0, x_2]} (R_1 + \sum_{i=2}^n x_i \cdot r_i^p).$$

The following lemma gives a threshold for the best reaction. Define $m' := m - \sum_{i>2} \alpha_i$ and $m_1' := m_1 - \sum_{i>2} \frac{x_i \alpha_i}{m_i + x_i}$.

Lemma 1. *Given pool 1's strategy* $\mathbf{x} = (x_2, \ldots, x_n)$ *and given reactions* $\boldsymbol{\alpha}_{-2} = (\alpha_3, \ldots, \alpha_n)$, *if* $x_2 \leq \frac{m_1' m_2}{m' - m_1}$, *then the best reaction of infiltrating miner 2 is* $\alpha_2^* = x_2$. *If* $x_2 > \frac{m_1' m_2}{m' - m_1}$, *then* $\alpha_2^* = 0$.

Proof. Define a function

$$\begin{aligned} f(\alpha_2) &:= R_1 + \sum_{i=2}^n x_i \cdot r_i^p \\ &= \frac{m_1 - \sum_i x_i}{m - \sum_i \alpha_i} + \sum_{i=2}^n \frac{x_i}{m_i + x_i} \cdot \frac{m_i + x_i - \alpha_i}{m - \sum_i \alpha_i} \end{aligned}$$

$$= \frac{1}{m - \sum_i \alpha_i} \left(m_1 - \sum_i x_i + \sum_{i=2}^{n} \frac{x_i(m_i + x_i - \alpha_i)}{m_i + x_i} \right)$$
$$= \frac{1}{m' - \alpha_2} \cdot \left(m_1' - \frac{x_2 \alpha_2}{m_2 + x_2} \right)$$
$$= \frac{1}{m_2 + x_2} \cdot \frac{m_1'(m_2 + x_2) - x_2\alpha_2}{m' - \alpha_2}.$$

Note that both m' and m_1' are independent of α_2 and can be regarded as constants. As the coefficient $\frac{1}{m_2+x_2}$ is a constant, by computing the derivative, $f(\alpha_2)$ is maximized as the lemma claims. □

We remark that when $x_2 = \frac{m_1' m_2}{m' - m_1}$, it has $u_2(\mathbf{x}, (0, \boldsymbol{\alpha}_{-2})) = u_2(\mathbf{x}, (x_2, \boldsymbol{\alpha}_{-2}))$, that is, both x_2 and 0 are the best reaction. In this case, we always assume that infiltrating miner 2 will not betray at all, i.e., $\alpha_2^* = x_2$.

3.2 Pool 1's Optimal Strategy

In the section, we study the optimal strategy of pool 1. First, we arbitrarily fix x_i and $\alpha_i \leq x_i$ for $i \geq 3$, and let x_2 be a variable. Provided that rational miner 2 reacts optimally as in Lemma 1, we study the optimal attacking power on pool 2: $x_2^* = \arg\max_{x_2} u_1(\mathbf{x}, \boldsymbol{\alpha})$. Let $x_2 > \frac{m_1' m_2}{m' - m_1}$ be an arbitrary value greater than the threshold, and thus miner 2's reaction is $\alpha_2 = 0$. The following lemma says that the utility of pool 1 when acting $x_2 > \frac{m_1' m_2}{m' - m_1}$ is less than that when acting $x_2 = 0$, under best reactions of miner 2. Therefore, it should always be $x_2^* \leq \frac{m_1' m_2}{m' - m_1}$.

Lemma 2. *For any fixed* $x_3, \ldots, x_n, \alpha_3, \ldots, \alpha_n$, *let* $\mathbf{x}' = (x_2', x_3, \cdots, x_n)$ *with* $x_2' > \frac{m_1' m_2}{m' - m_1}$, $\mathbf{x}'' = (0, x_3, \cdots, x_n)$, *and* $\boldsymbol{\alpha} = (0, \alpha_3, \cdots, \alpha_n)$. *Then* $u_1(\mathbf{x}', \boldsymbol{\alpha}) < u_1(\mathbf{x}'', \boldsymbol{\alpha})$. *In addition, when* $x_2' = \frac{m_1' m_2}{m' - m_1}$, *then* $u_1(\mathbf{x}', \boldsymbol{\alpha}) = u_1(\mathbf{x}'', \boldsymbol{\alpha})$.

Proof. Recall that the utility of pool 1 is defined as

$$u_1(\mathbf{x}, \boldsymbol{\alpha}) = \left[m_1 - \sum_{i=2}^{n} (x_i - \alpha_i) \right] \cdot r_1^p(\mathbf{x}, \boldsymbol{\alpha}).$$

Because $m_1 - x_2' - \sum_{i=3}^{n}(x_i - \alpha_i) < m_1 - \sum_{i=3}^{n}(x_i - \alpha_i)$, it suffices to prove that $r_1^p(\mathbf{x}', \boldsymbol{\alpha}) \leq r_1^p(\mathbf{x}'', \boldsymbol{\alpha})$, which is equivalent to

$$R_1(\mathbf{x}', \boldsymbol{\alpha}) + x_2' \cdot r_2^p(\mathbf{x}', \boldsymbol{\alpha}) + \sum_{i=3}^{n} x_i \cdot r_i^p(\mathbf{x}', \boldsymbol{\alpha})$$
$$\leq R_1(\mathbf{x}'', \boldsymbol{\alpha}) + \sum_{i=3}^{n} x_i \cdot r_i^p(\mathbf{x}'', \boldsymbol{\alpha}).$$

We have

$$
\begin{aligned}
&R_1(\mathbf{x}',\boldsymbol{\alpha}) + x_2' \cdot r_2^p(\mathbf{x}',\boldsymbol{\alpha}) + \sum_{i=3}^{n} x_i \cdot r_i^p(\mathbf{x}',\boldsymbol{\alpha}) \\
&= \frac{m_1 - \sum_{i=3}^{n} x_i - x_2'}{m - \sum_i \alpha_i} + x_2' \cdot \frac{m_2 + x_2' - 0}{(m_2 + x_2')\cdot(m - \sum_j \alpha_j)} \\
&\quad + \sum_{i=3}^{n} x_i \cdot \frac{m_i + x_i - \alpha_i}{(m_i + x_i)\cdot(m - \sum_j \alpha_j)} \\
&= \frac{m_1 - \sum_{i=3}^{n} x_i}{m - \sum_i \alpha_i} + \sum_{i=3}^{n} x_i \cdot \frac{m_i + x_i - \alpha_i}{(m_i + x_i)\cdot(m - \sum_j \alpha_j)} \\
&= R_1(\mathbf{x}'',\boldsymbol{\alpha}) + \sum_{i=3}^{n} x_i \cdot r_i^p(\mathbf{x}'',\boldsymbol{\alpha}),
\end{aligned}
$$

as desired. □

Recall that $\mathbf{x}^* = \arg\max_{\mathbf{x}} u_1(\mathbf{x}, \boldsymbol{\alpha}_{\mathbf{x}})$ is the optimal strategy of pool 1, where $\boldsymbol{\alpha}_{\mathbf{x}}$ is the best reaction profile of infiltrating miners w.r.t. pool 1's strategy $\mathbf{x}$. Note that the utility of pool 1 without any infiltration is $u_1(\mathbf{0},\mathbf{0}) = \frac{m_1}{m}$. Because

$$
u_1(x_r, x_r) - \frac{m_1}{m} = \frac{x_r \cdot [(m_1 - m)x_r + m_1 m_2]}{m(m - x_r)(m_2 + x_r)} \geq 0,
$$

the infiltration always brings benefits.

Now, we are able to prove the main theorem in this section.

Theorem 1. *The optimal strategy of pool 1* $\mathbf{x}^* = (x_2^*, \ldots, x_n^*)$ *is a solution of the equation system: for all* $i = 2, \ldots, n$,

$$
\begin{aligned}
0 =& (m_1 - m + \sum_{k \neq i}^{n} \frac{m_k \cdot x_k^*}{m_k + x_k^*})(m_i + x_i^*)^2 + m_i^2(m - \sum_{j=2}^{n} x_j^*) \\
&+ m_i x_i^*(m_i + x_i^*).
\end{aligned}
$$

In addition the infiltrating miners will not betray pool 1 at all.

Proof. Let $\boldsymbol{\alpha}_{\mathbf{x}^*} = (\alpha_2^*, \ldots, \alpha_n^*)$ be the optimal reactions of infiltrating miners w.r.t. $\mathbf{x}^*$. By Lemma 2, pool 1's optimal strategy $\mathbf{x}^*$ ensures that every miner is completely loyal, i.e., $\alpha_i^* = x_i^*$ for all $i \geq 2$. Substituting them to the utility function, we have

$$
\begin{aligned}
u_1(\mathbf{x},\boldsymbol{\alpha}) &= [m_1 - \sum_{i=2}^{n}(x_i - \alpha_i)] \cdot r_1^p(\mathbf{x},\boldsymbol{\alpha}) \\
&= R_1(\mathbf{x},\boldsymbol{\alpha}) + \sum_{i=2}^{n} x_i \cdot r_i^p(\mathbf{x},\boldsymbol{\alpha})
\end{aligned}
$$

$$= \frac{m_1 - \sum_{i=2}^n x_i}{m - \sum_{i=2}^n x_i} + \sum_{i=2}^n \frac{x_i \cdot m_i}{(m_i + x_i)(m - \sum_i x_i)}.$$

The derivative of $u_1(\mathbf{x}, \boldsymbol{\alpha})$ for each x_i $(i \geq 2)$ is

$$\frac{\partial u_1}{\partial x_i} = \frac{m_1 - m}{(m - \sum_{j=2}^n x_j)^2} + m_i \cdot \frac{m_i(m - \sum_{j=2}^n x_j) + x_i(m_i + x_i)}{(m_i + x_i)^2(m - \sum_{j=2}^n x_j)^2}$$
$$+ \sum_{k \neq i}^n \frac{m_k \cdot x_k}{m_k + x_k} \cdot \frac{1}{(m - \sum_{j=2}^n x_j)^2}$$
$$= \frac{P}{(m_i + x_i)^2(m - \sum_{j=2}^n x_j)^2},$$

where

$$P = (m_1 - m + \sum_{k \neq i}^n \frac{m_k \cdot x_k}{m_k + x_k})(m_i + x_i)^2 + m_i^2(m - \sum_{j=2}^n x_j) + m_i x_i(m_i + x_i).$$

W.l.o.g., we only consider $i = 2$. It suffices to show that x_2^* can never be a boundary point (i.e., 0 or $\frac{m_1' m_2}{m' - m_1}$), which implies that $P = 0$. By Lemma 2, the utility of pool 1 is the same when $x_2 = 0$ and $x_2 = \frac{m_1' m_2}{m' - m_1}$, fixing all other variables. Moreover, when $x_2 = 0$, we have

$$P = m_i^2(m_1 - \sum_{k \neq i}^n \frac{x_k^2}{m_k + x_k}) > 0,$$

and thus x_2^* must be in the interval $(0, \frac{m_1' m_2}{m' - m_1})$. Since $u_1(\mathbf{x}, \boldsymbol{\alpha})$ is differentiable at every point within $(0, \frac{m_1' m_2}{m' - m_1})$, we have $\frac{\partial u_1(\mathbf{x}^*, \boldsymbol{\alpha}_{\mathbf{x}^*})}{\partial x_2} = 0$, as desired. □

As a corollary we give an explicit expression when there are two mining pools.

Corollary 1. *When $n = 2$, if $m > m_1 + m_2$, then the optimal strategy of pool 1 is*

$$x_2^* = \frac{m_2\sqrt{m^2 - m_1 m_2 - m m_1} - (m - m_1)\, m_2}{m - m_2 - m_1}.$$

If $m = m_1 + m_2$, then $x_2^* = \frac{m_1 m_2}{2(m - m_1)}$. In both cases, the infiltrating miner will not betray pool 1 at all.

Proof. The claim simply follows from Theorem 1. In both cases, we have $x_2^* \in (0, \frac{m_1 m_2}{m - m_1})$. □

Table 1. Experiments for computing the optimal strategy in the BR Model.

Experiments	m_1	m_2	m_3	m_4	m_5	$\mathbf{x}^*$	u_1^*	u_1^*/u_1^{non}
(1)	0.100	0.120	0.190	0.260	0.330	(0.007, 0.011, 0.014, 0.018)	0.102	1.025
(2)	0.120	0.115	0.185	0.255	0.325	(0.008, 0.013, 0.017, 0.022)	0.124	1.030
(3)	0.140	0.110	0.180	0.250	0.320	(0.009, 0.015, 0.020, 0.026)	0.145	1.035
(4)	0.160	0.105	0.175	0.245	0.315	(0.010, 0.017, 0.023, 0.030)	0.166	1.040
(5)	0.180	0.100	0.170	0.240	0.310	(0.011, 0.019, 0.026, 0.034)	0.188	1.045
(6)	0.200	0.095	0.165	0.235	0.305	(0.012, 0.021, 0.029, 0.038)	0.210	1.049
(7)	0.220	0.090	0.160	0.230	0.300	(0.013, 0.023, 0.032, 0.042)	0.232	1.054
(8)	0.240	0.085	0.155	0.225	0.295	(0.013, 0.024, 0.036, 0.047)	0.254	1.059
(9)	0.260	0.080	0.150	0.220	0.290	(0.014, 0.026, 0.039, 0.051)	0.277	1.064
(10)	0.280	0.075	0.145	0.215	0.285	(0.015, 0.028, 0.042, 0.055)	0.299	1.068
(11)	0.300	0.070	0.140	0.210	0.280	(0.015, 0.030, 0.045, 0.060)	0.322	1.073
(12)	0.320	0.065	0.135	0.205	0.275	(0.015, 0.032, 0.048, 0.065)	0.345	1.077

3.3 Simulations

In the simulation, we do experiments to investigate the strategic behavior of pool 1. It illustrates what fraction of the mining power is used for infiltrating, and how many benefits are brought by infiltrating. Table 1 presents 12 typical examples, and more examples are given in Appendix.

In all of the examples, suppose there are 5 mining pools, and set the total mining power $m = m_1 + \cdots + m_5 = 1$. For different combinations of m_i, we compute the optimal strategy $\mathbf{x}^*$ of pool 1, and the corresponding optimal utility u_1^*. If pool 1 does not attack (i.e., $\mathbf{x} = \mathbf{0}$), clearly the utility is $u_1^{non} = \frac{m_1}{m} = m_1$. We use the ratio $\frac{u_1^*}{u_1^{non}}$ to measure the benefits that attacking can bring.

Our experiment results show that, about 50% mining power of pool 1 is used for attacking in an optimal strategy, which brings a less than 10% increase of utility.

4 Betray-and-Not-Return Model

In this section, we discuss the Betray-and-Not-Return model, in which the infiltrating miners who betray pool 1 will not bring any reward share back to pool 1. We first study in Sect. 4.1 the strategic behavior of pool 1 when it can only attack a single pool, and then in Sect. 4.2 we implement experiments to show the best reaction of infiltrating miners in the general cases.

4.1 Attacking a Single Pool

We start from the two-pool case, i.e., $n = 2$, and thus pool 1 can only attack pool 2. To simplify the expression, we denote x_2 as x and α_2 as α. This case can also be generalized to the case where pool 1 is able to attack only a single pool, though there may be multiple $n \geq 3$ pools.

Then the utility function of pool 1 is $u_1(x, \alpha) = (m_1 - x + \alpha) \cdot r_1^d(x, \alpha)$, and the utility function of infiltrating miner 2 is

$$u_2(x, \alpha) = (x - \alpha) \cdot r_2^d(x, \alpha) + \alpha \cdot r_1^d(x, \alpha).$$

As in Sect. 3, we denote by α_x the best reaction of infiltrating miner 2 given pool 1's strategy x, and denote by x^* the optimal strategy of pool 1 given that the infiltrating miner reacts optimally. In the following we first give a threshold for the best reaction of miner 2.

Lemma 3. *Given pool 1's strategy x, if $x \leq \frac{m_1 m_2}{m - m_1}$, then the best reaction is $\alpha_x = x$, otherwise $\alpha_x = 0$. In addition, if $x = \frac{m_1 m_2}{m - m_1}$, then both x and 0 are the best reaction.*

Proof. The utility function of the infiltrating miner is

$$\begin{aligned} u_2(x, \alpha) &= (x - \alpha) \cdot r_2^d(x, \alpha) + \alpha \cdot r_1^d(x, \alpha) \\ &= \frac{(m_1' - x)\alpha^2 + x(m_2' - m_1')\alpha + x \cdot m_1' \cdot m_2'}{(m_1' + \alpha)(m - \alpha)m_2'}, \end{aligned}$$

where $m_1' = m_1 - x$ and $m_2' = m_2 + x$.

Fixing x, we regard $u_2(x, \alpha)$ as a single-variable function $u_2(\alpha)$. The derivative of u_2 is

$$u_2'(\alpha) = \frac{f(\alpha)}{[\alpha^2 + (m_1' - m)\alpha - m \cdot m_1'] \cdot m_2'},$$

where $f(\alpha) = [(m_2' - m)x + m_1'(m - m_1')]\alpha^2 + [2m_1'(m_2' - m)x + 2mm_1'^2]\alpha + (m_2' - m)x \cdot m_1'^2$.

Claim 1. *When $x \geq \frac{m_1 m_2}{m - m_1}$, $u_2(0) \geq u_2(\alpha)$ for any $\alpha \in [0, x]$.*

Proof. Note that $u_2(0) = \frac{x}{m}$. For any $\alpha \in [0, x]$, using notation $\propto$ referred to as "directly proportional to", we have

$$\begin{aligned} u_2(\alpha) - u_2(0) \propto\ & \alpha(x^2 - 2mx + m_2 x + mm_1) - x^3 \\ & + x^2(m + m_1 - m_2) + x(m_1 m_2 - mm_1). \end{aligned}$$

Define function $h(x) = x^2 - 2mx + m_2 x + mm_1$. We discuss three cases as follows.

Case 1. When $h(x) = 0$, we have

$$u_2(\alpha) - u_2(0) \propto q(x),$$

where $q(x) = -x^2 + x(m + m_1 - m_2) + m_1 m_2 - mm_1$. Compute the derivative of $q(x)$:

$$q'(x) = -2x + m + m_1 - m_2 \geq 2m_1 - 2x.$$

It follows that, in interval $[0, m_1]$, function $q(x)$ is increasing, and the maximum is achieved by $q(m_1) = -m_1^2 + m_1(m + m_1 - m_2) + m_1 m_2 - mm_1 = 0$. Therefore, $u_2(\alpha) - u_2(0) \leq 0$.

Case 2. When $h(x) > 0$, we have

$$\begin{aligned}
&u_2(\alpha) - u_2(0)\\
&\propto\ \alpha(x^2 - 2mx + m_2x + mm_1) - x^3 + x^2(m + m_1 - m_2)\\
&\quad + x(m_1m_2 - mm_1)\\
&\le\ x(x^2 - 2mx + m_2x + mm_1) - x^3 + x^2(m + m_1 - m_2)\\
&\quad + x(m_1m_2 - mm_1)\\
&=\ [(m_1 - m)x + m_1m_2] \cdot x\\
&\le\ \left[(m_1 - m)\frac{m_1m_2}{m - m_1} + m_1m_2\right] \cdot x \qquad (3)\\
&=\ 0,
\end{aligned}$$

where Eq. (3) is because $m_1 - m < 0$ and $x \ge \frac{m_1m_2}{m-m_1}$.

Case 3. When $h(x) < 0$, we have

$$\begin{aligned}
&u_2(\alpha) - u_2(0)\\
&\propto\ \alpha(x^2 - 2mx + m_2x + mm_1) - x^3\\
&+ x^2(m + m_1 - m_2) + x(m_1m_2 - mm_1)\\
&\le\ q(x) \cdot x\\
&\le 0,
\end{aligned}$$

which establishes the proof.

Claim 2. *When $x \le \frac{m_1m_2}{m-m_1}$, $u_2(x) \ge u_2(\alpha)$ for any $\alpha \in [0, x]$.*

Proof. First, by a simple calculation, we have $\frac{m_1m_2}{m-m_1} \le \frac{m_1(m-m_1)}{2m-2m_1-m_2}$. We rewrite $f(\alpha)$ as

$$f(\alpha) = A \cdot \alpha^2 + B \cdot \alpha + C,$$

where

$$\begin{cases} A = (2m_1 + m_2 - 2m)x + m_1(m - m_1),\\ B = -2x^3 + (4m + 2m_1 - 2m_2)x^2\\ \qquad +2m_1(m_2 - 3m)x + 2mm_1^2\\ C = (m_2 - m + x)(m_1 - x)^2x. \end{cases} \qquad (4)$$

Note that when $x \le \frac{m_1m_2}{m-m_1} < \frac{m_1(m-m_1)}{2m-2m_1-m_2}$, $A > 0$, and $f(\alpha)$ is a quadratic function with a parabola opening upward; when $x = \frac{m_1m_2}{m-m_1} = \frac{m_1(m-m_1)}{2m-2m_1-m_2}$, $A > 0$, and $f(\alpha)$ is linear. Combining with $f(0) \le 0$, we can easily conclude that when $x \le \frac{m_1m_2}{m-m_1}$, either $u_2(0)$ or $u_2(x)$ is the maximum of function $u_2(\alpha)$ in domain $[0, x]$. So it suffices to prove that $u_2(x) \ge u_2(0)$, when $x < \frac{m_1m_2}{m-m_1}$.

Substituting $\alpha = x$, we have

$$u_2(x) - u_2(0)$$

$$
\begin{aligned}
&\propto x(x^2 - 2mx + m_2x + mm_1) - x^3 \\
&+ x^2(m + m_1 - m_2) + x(m_1m_2 - mm_1) \\
&= [(m_1 - m)x + m_1m_2] \cdot x \\
&\geq \left[(m_1 - m)\frac{m_1m_2}{m - m_1} + m_1m_2\right] \cdot x \\
&= 0,
\end{aligned}
$$

completing the proof.

Combining Claims 1 and 2, we obtain the lemma. □

Now we can compute the optimal strategy of pool 1.

Theorem 2. *When $n = 2$, if $m_1 + m_2 < m$, the optimal strategy for pool 1 is*

$$
x^* = \frac{m_2\sqrt{-m_1m_2 - mm_1 + m^2} - (m - m_1)m_2}{m - m_1 - m_2}.
$$

If $m_1 + m_2 - m = 0$, then $x^* = \frac{m_1m_2}{2(m-m_1)}$. In both cases, the infiltrating miner will not betray pool 1 at all.

Proof. Recall that $u_1(x, \alpha) = (m_1 - x + \alpha) \cdot r_1(x, \alpha)$, and

$$
r_1(x, \alpha) = \frac{(m_1 - x)(m_2 + x) + \alpha(m_2 + x - \alpha)}{(m_1 - x + \alpha)(m - \alpha)(m_2 + x)}.
$$

When $x > \frac{m_1m_2}{m-m_1}$, by Lemma 3, the utility of pool 1 is

$$
u_1(x, 0) = (m_1 - x) \cdot \frac{(m_1 - x)(m_2 + x)}{(m_1 - x)m(m_2 + x)} = \frac{m_1 - x}{m} < u_1(0, 0),
$$

implying that pool 1 will never set $x > \frac{m_1m_2}{m-m_1}$.

When $x \leq \frac{m_1m_2}{m-m_1}$, by Lemma 3, the utility of pool 1 is

$$
u_1(x, x) = m_1 \cdot r_1(x, x) = m_1 \cdot \frac{(m_1 - x)(m_2 + x) + m_2x}{m_1(m - x)(m_2 + x)}.
$$

The derivative is

$$
\frac{\partial u_1}{\partial x} = \frac{(m_1 + m_2 - m)x^2 + (2m_1m_2 - 2mm_2)x + m_1m_2^2}{(x + m_2)^2 \cdot (x - m)^2}.
$$

If $m_1 + m_2 - m = 0$, then it is easy to see that $x^* = \frac{m_1m_2}{2(m-m_1)}$. If $m_1 + m_2 - m < 0$, the numerator of $\frac{\partial U_1}{\partial x}$ is a quadratic function w.r.t. x, with two roots p_1, p_2:

$$
p_1 = -\frac{m_2\sqrt{-m_1m_2 - mm_1 + m^2} + (m - m_1)m_2}{m - m_1 - m_2} < 0,
$$

$$
0 < p_2 = \frac{m_2\sqrt{-m_1m_2 - mm_1 + m^2} - (m - m_1)m_2}{m - m_1 - m_2} \leq \frac{m_1m_2}{m - m_1}.
$$

Therefore, we have $x^* = p_2$. □

4.2 Attacking Multiple Pools and Simulations

The BNR model in the general case is much more complicated. First, we prove that if an infiltrating miner i has completely betrayed (i.e., $\alpha_i = 0$), then the utility of pool 1 will never be better than that when it does not infiltrate pool i.

Lemma 4. *For any fixed $x_3, \ldots, x_n, \alpha_3, \ldots, \alpha_n$, let $\mathbf{x}' = (x_2', x_3, \cdots, x_n)$ with $x_2' > 0$, $\mathbf{x}'' = (0, x_3, \ldots, x_n)$, and $\boldsymbol{\alpha} = (0, \alpha_3, \ldots, \alpha_n)$. Then $u_1(\mathbf{x}', \boldsymbol{\alpha}) < u_1(\mathbf{x}'', \boldsymbol{\alpha})$.*

Proof. Recall that the utility of pool 1 is defined as

$$u_1(\mathbf{x}, \alpha) = [m_1 - \sum_{i=2}^{n}(x_i - \alpha_i)] \cdot r_1^d(\mathbf{x}, \alpha)$$
$$= \frac{m_1 - \sum_{j=2}^{n} x_j + \sum_{j=2}^{n} \frac{\alpha_j \cdot (m_j + x_j - \alpha_j)}{m_j + x_j}}{m - \sum_{j=2}^{n} \alpha_j}.$$

So we have

$$u_1(\mathbf{x}', \alpha) = \frac{m_1 - \sum_{j>2}^{n} x_j + \sum_{j>2}^{n} \frac{\alpha_j \cdot (m_j + x_j - \alpha_j)}{m_j + x_j} - x_2}{m - \sum_{j>2}^{n} \alpha_j},$$

$$u_1(\mathbf{x}'', \alpha) = \frac{m_1 - \sum_{j>2}^{n} x_j + \sum_{j>2}^{n} \frac{\alpha_j \cdot (m_j + x_j - \alpha_j)}{m_j + x_j}}{m - \sum_{j>2}^{n} \alpha_j}.$$

Clearly, the lemma is proved. □

Based on the results on the BR model and Lemma 3 for the two-pool case, a natural conjecture is that, the best reaction α_i of any infiltrating miner i is binary (i.e., either $\alpha_i = x_i$ or $\alpha_i = 0$), and there is a threshold for pool 1's attacking strategy x_i on pool i, such that when x_i is smaller then miner i would be completely loyal; when x_i is larger then it would completely betray.

Conjecture 1. For any fixed $x_3, \ldots, x_n, \alpha_3, \ldots, \alpha_n$, there exists a threshold t for x_2: if $x_2 \leq t$, then the best reaction of infiltrating miner i is $\alpha_i = x$; if $x > t$, then $\alpha_i = 0$.

Though we are unable to prove it theoretically, we provide simulation results to verify it. In our experiments, fixing $m_1, \ldots, m_n, x_3, \ldots, x_n, \alpha_3, \ldots, \alpha_n$ arbitrarily, we take x_2 as a variable, and increase the value of x_2. For each value of x_2, we compute the best reaction α_2 of infiltrating miner 2. The experiments results show that such a threshold truly exists in every example, and thus support the conjecture. Moreover, on average, the threshold in the BNR model is at least 20% smaller than that in the BR model, which implies that the infiltrating miners in the BNR model are more likely to betray their original pool.

Table 2 shows three examples in which we fix $m_1, \ldots, m_5, x_3, \ldots, x_5, \alpha_3, \ldots, \alpha_5$, and let x_2 be a variable increasing from 0.001 within a step length of 0.002. For each value of x_2, we compute a corresponding best reaction α_2^* of miner 2. A rough threshold t is marked in bold for each example: when $x_2 < t$, $\alpha_2^* = x_2$; when $x_2 > t$, $\alpha_2^* = 0$. For the three examples, the thresholds in this BNR model are $t \approx 0.009, 0.005, 0.003$, respectively, while the threshold for these examples in the BR Model is $t^{BR} = 0.014, 0.025, 0.032$, respectively. Full examples can be found in Table 4 in Appendix. We conclude that, for every example, the threshold in the BNR model is much smaller than that in the BR model, which implies that the pool in BNR model is less likely to attack other pools.

Table 2. Experiments for computing miner 2's best reaction in the BNR model, where $\mathbf{x} = (x_2, x_3 = 0.1m_1, x_4 = 0.12m_1, x_5 = 0.15m_1)$ and $\boldsymbol{\alpha} = (\alpha_2, \alpha_3 = 0, \alpha_4 = 0.5x_4, \alpha_5 = x_5)$.

Experiments		(1)	(2)	(3)	(4)	(5)	(6)	(7)	(8)	(9)	(10)
$m_1 = 0.100, m_2 = 0.120,$	x_2	0.001	0.003	0.005	0.007	0.009	0.011	0.013	0.015	...	0.063
$m_3 = 0.190, m_4 = 0.260,$	α_2^*	0.001	0.003	0.005	0.007	**0.009**	**0.000**	0.000	0.000	0.000	0.000
$m_5 = 0.330, t^{BR} = 0.014$	u_1	0.085	0.085	0.085	0.085	0.085	0.074	0.072	0.070	...	0.021
$m_1 = 0.200, m_2 = 0.095,$	x_2	0.001	0.003	0.005	0.007	0.009	0.011	0.013	0.015	...	0.095
$m_3 = 0.165, m_4 = 0.235,$	α_2^*	0.001	0.003	**0.005**	**0.000**	0.000	0.000	0.000	0.000	0.000	0.000
$m_5 = 0.305, t^{BR} = 0.025$	u_1	0.172	0.172	0.173	0.165	0.163	0.161	0.158	0.156	...	0.073
$m_1 = 0.300, m_2 = 0.070,$	x_2	0.001	0.003	0.005	0.007	0.009	0.011	0.013	0.015	...	0.069
$m_3 = 0.140, m_4 = 0.210,$	α_2^*	0.001	**0.003**	**0.000**	0.000	0.000	0.000	0.000	0.000	0.000	0.000
$m_5 = 0.280, t^{BR} = 0.032$	u_1	0.261	0.262	0.256	0.253	0.251	0.249	0.247	0.245	...	0.187

5 Conclusion

In this paper, for the first time, we studied pool block withholding attack in Bitcoin where infiltrating miners are strategic. For two betrayal models depending on whether betraying miners return to the original pool or not, we investigated the optimal behaviors for both miners and the malicious pool, theoretically and empirically. There are many interesting future directions that are related to our work. For example, since we only considered the case of one pool attacking others, it would be interesting to take the game theoretic view similar to [8] and [1], where multiple pools attack each other, and analyze the existence of Nash equilibrium and the price of anarchy.

A Appendix

In the Appendix, we present the full experiment results.

A.1 Experiments in the BR Model

In the simulations, we do experiments to investigate the strategic behavior of pool 1. It illustrates how many fraction of the mining power is used for infiltrating, and how many benefits are brought by infiltrating. In all of the examples (see Table 3), suppose there are 5 mining pools, and set the total mining power

$m = m_1 + \cdots + m_5 = 1$. For different combinations of m_i, we compute the optimal strategy $\mathbf{x}^*$ of pool 1, and the corresponding optimal utility u_1^*. If pool 1 does not attack (i.e., $\mathbf{x} = \mathbf{0}$), clearly the utility is $u_1^{non} = \frac{m_1}{m} = m_1$. We use the ratio $\frac{u_1^*}{u_1^{non}}$ to measure the benefits that attacking can bring.

Our experiment results show that, about 50% mining power of pool 1 is used for attacking in an optimal strategy, which will bring a less than 10% increase of utility.

Table 3. Experiments for computing the optimal strategy in the BR Model.

m_1	m_2	m_3	m_4	m_5	$\mathbf{x}^*$	u_1^*	u_1^*/u_1^{non}
0.060	0.110	0.200	0.280	0.350	(0.004, 0.006, 0.009, 0.011)	0.061	1.015
0.072	0.110	0.196	0.276	0.346	(0.004, 0.008, 0.011, 0.013)	0.073	1.018
0.084	0.110	0.192	0.272	0.342	(0.005, 0.009, 0.012, 0.016)	0.086	1.021
0.096	0.110	0.188	0.268	0.338	(0.006, 0.010, 0.014, 0.018)	0.098	1.024
0.108	0.110	0.184	0.264	0.334	(0.007, 0.011, 0.016, 0.020)	0.111	1.027
0.120	0.110	0.180	0.260	0.330	(0.008, 0.012, 0.018, 0.023)	0.124	1.030
0.132	0.110	0.176	0.256	0.326	(0.008, 0.013, 0.019, 0.025)	0.136	1.033
0.144	0.110	0.172	0.252	0.322	(0.009, 0.014, 0.021, 0.027)	0.149	1.036
0.156	0.110	0.168	0.248	0.318	(0.010, 0.016, 0.023, 0.029)	0.162	1.039
0.168	0.110	0.164	0.244	0.314	(0.011, 0.017, 0.025, 0.032)	0.175	1.042
0.180	0.110	0.160	0.240	0.310	(0.012, 0.018, 0.026, 0.034)	0.188	1.045
0.192	0.110	0.156	0.236	0.306	(0.013, 0.019, 0.028, 0.036)	0.201	1.047
0.204	0.110	0.152	0.232	0.302	(0.014, 0.019, 0.030, 0.039)	0.214	1.050
0.216	0.110	0.148	0.228	0.298	(0.015, 0.020, 0.031, 0.041)	0.227	1.053
0.228	0.110	0.144	0.224	0.294	(0.016, 0.021, 0.033, 0.043)	0.241	1.056
0.240	0.110	0.140	0.220	0.290	(0.017, 0.022, 0.035, 0.046)	0.254	1.059
0.252	0.110	0.136	0.216	0.286	(0.019, 0.023, 0.036, 0.048)	0.268	1.062
0.264	0.110	0.132	0.212	0.282	(0.020, 0.024, 0.038, 0.051)	0.281	1.064
0.276	0.110	0.128	0.208	0.278	(0.021, 0.024, 0.040, 0.053)	0.295	1.067
0.288	0.110	0.124	0.204	0.274	(0.022, 0.025, 0.041, 0.055)	0.308	1.070
0.300	0.110	0.120	0.200	0.270	(0.024, 0.026, 0.043, 0.058)	0.322	1.073
0.312	0.110	0.116	0.196	0.266	(0.025, 0.026, 0.044, 0.060)	0.336	1.075
0.324	0.110	0.112	0.192	0.262	(0.026, 0.027, 0.046, 0.063)	0.349	1.078
0.336	0.110	0.108	0.188	0.258	(0.028, 0.027, 0.048, 0.065)	0.363	1.081
0.348	0.110	0.104	0.184	0.254	(0.029, 0.028, 0.049, 0.068)	0.377	1.083
0.360	0.110	0.100	0.180	0.250	(0.031, 0.028, 0.051, 0.070)	0.391	1.086

A.2 Experiments in the BNR Model

In our experiments for the BNR model, arbitrarily fixing $m_1, \ldots, m_n, x_3, \ldots, x_n$, $\alpha_3, \ldots, \alpha_n$, we take x_2 as a variable, and increase the value of x_2. For each value of x_2, we compute the best reaction α_2 of infiltrating miner 2. The experiments results show that such a threshold truly exists in every example, and thus support the conjecture. Moreover, on average, the threshold in the BNR model is at least 20% smaller than that in the BR model, which implies that the infiltrating miners in the BNR model are more likely to betray their original pool.

In Table 4, we fix $m_1, \ldots, m_5, x_3, \ldots, x_5, \alpha_3, \ldots, \alpha_5$, and let x_2 be a variable increasing from 0.001 within a step length of 0.002. For each value of x_2, we compute a corresponding best reaction α_2^* of miner 2. A rough threshold t is given for each example: when $x_2 < t$, $\alpha_2^* = x_2$; when $x_2 > t$, $\alpha_2^* = 0$. The threshold for these examples in the BR model is denoted by t^{BR}. We conclude that, for every example, the threshold in the BNR model is much smaller than that in the BR model, which implies that the pool in BNR model is less likely to attack other pools.

Table 4. Experiments for computing miner 2's best reaction in the BNR model, where $\mathbf{x} = (x_2, x_3 = 0.1m_1, x_4 = 0.12m_1, x_5 = 0.15m_1)$ and $\boldsymbol{\alpha} = (\alpha_2, \alpha_3 = 0, \alpha_4 = 0.5x_4, \alpha_5 = x_5)$.

Experiments		(1)	(2)	(3)	(4)	(5)	(6)	(7)	(8)	(9)	(10)
$m_1 = 0.100, m_2 = 0.120,$	x_2	0.001	0.003	0.005	0.007	0.009	0.011	0.013	0.015	...	0.063
$m_3 = 0.190, m_4 = 0.260,$	α_2^*	0.001	0.003	0.005	0.007	0.009	0.000	0.000	0.000	0.000	0.000
$m_5 = 0.330, t^{BR} = 0.014$	u_1	0.085	0.085	0.085	0.085	0.085	0.074	0.072	0.070	...	0.021
$m_1 = 0.120, m_2 = 0.115,$	x_2	0.001	0.003	0.005	0.007	0.009	0.011	0.013	0.015	...	0.075
$m_3 = 0.185, m_4 = 0.255,$	α_2^*	0.001	0.003	0.005	0.007	0.009	0.000	0.000	0.000	0.000	0.000
$m_5 = 0.325, t^{BR} = 0.016$	u_1	0.102	0.102	0.103	0.103	0.103	0.091	0.089	0.087	...	0.025
$m_1 = 0.140, m_2 = 0.110,$	x_2	0.001	0.003	0.005	0.007	0.009	0.011	0.013	0.015	...	0.089
$m_3 = 0.180, m_4 = 0.250,$	α_2^*	0.001	0.003	0.005	0.007	0.009	0.000	0.000	0.000	0.000	0.000
$m_5 = 0.320, t^{BR} = 0.018$	u_1	0.120	0.120	0.120	0.120	0.120	0.108	0.106	0.104	...	0.028
$m_1 = 0.160, m_2 = 0.105,$	x_2	0.001	0.003	0.005	0.007	0.009	0.011	0.013	0.015	...	0.101
$m_3 = 0.175, m_4 = 0.245,$	α_2^*	0.001	0.003	0.005	0.007	0.000	0.000	0.000	0.000	0.000	0.000
$m_5 = 0.315, t^{BR} = 0.021$	u_1	0.137	0.137	0.137	0.137	0.128	0.126	0.124	0.121	...	0.032
$m_1 = 0.180, m_2 = 0.100,$	x_2	0.001	0.003	0.005	0.007	0.009	0.011	0.013	0.015	...	0.099
$m_3 = 0.170, m_4 = 0.240,$	α_2^*	0.001	0.003	0.005	0.007	0.000	0.000	0.000	0.000	0.000	0.000
$m_5 = 0.310, t^{BR} = 0.023$	u_1	0.155	0.155	0.155	0.155	0.145	0.143	0.141	0.139	...	0.052
$m_1 = 0.200, m_2 = 0.095,$	x_2	0.001	0.003	0.005	0.007	0.009	0.011	0.013	0.015	...	0.095
$m_3 = 0.165, m_4 = 0.235,$	α_2^*	0.001	0.003	0.005	0.000	0.000	0.000	0.000	0.000	0.000	0.000
$m_5 = 0.305, t^{BR} = 0.025$	u_1	0.172	0.172	0.173	0.165	0.163	0.161	0.158	0.156	...	0.073
$m_1 = 0.220, m_2 = 0.090,$	x_2	0.001	0.003	0.005	0.007	0.009	0.011	0.013	0.015	...	0.089
$m_3 = 0.160, m_4 = 0.230,$	α_2^*	0.001	0.003	0.005	0.000	0.000	0.000	0.000	0.000	0.000	0.000
$m_5 = 0.300, t^{BR} = 0.026$	u_1	0.190	0.190	0.190	0.182	0.180	0.178	0.176	0.174	...	0.096
$m_1 = 0.240, m_2 = 0.085,$	x_2	0.001	0.003	0.005	0.007	0.009	0.011	0.013	0.015	...	0.085
$m_3 = 0.155, m_4 = 0.225,$	α_2^*	0.001	0.003	0.005	0.000	0.000	0.000	0.000	0.000	0.000	0.000
$m_5 = 0.295, t^{BR} = 0.028$	u_1	0.208	0.208	0.208	0.200	0.198	0.196	0.194	0.192	...	0.118

(continued)

Table 4. (*continued*)

Experiments		(1)	(2)	(3)	(4)	(5)	(6)	(7)	(8)	(9)	(10)
$m_1 = 0.260, m_2 = 0.080,$	x_2	0.001	0.003	0.005	0.007	0.009	0.011	0.013	0.015	...	0.079
$m_3 = 0.150, m_4 = 0.220,$	α_2^*	0.001	0.003	0.000	0.000	0.000	0.000	0.000	0.000	0.000	0.000
$m_5 = 0.290, t^{BR} = 0.030$	u_1	0.225	0.226	0.220	0.218	0.216	0.214	0.211	0.209	...	0.142
$m_1 = 0.280, m_2 = 0.075,$	x_2	0.001	0.003	0.005	0.007	0.009	0.011	0.013	0.015	...	0.075
$m_3 = 0.145, m_4 = 0.215,$	α_2^*	0.001	0.003	0.000	0.000	0.000	0.000	0.000	0.000	0.000	0.000
$m_5 = 0.285, t^{BR} = 0.031$	u_1	0.243	0.244	0.238	0.236	0.233	0.231	0.229	0.227	...	0.163
$m_1 = 0.300, m_2 = 0.070,$	x_2	0.001	0.003	0.005	0.007	0.009	0.011	0.013	0.015	...	0.069
$m_3 = 0.140, m_4 = 0.210,$	α_2^*	0.001	0.003	0.000	0.000	0.000	0.000	0.000	0.000	0.000	0.000
$m_5 = 0.280, t^{BR} = 0.032$	u_1	0.261	0.262	0.256	0.253	0.251	0.249	0.247	0.245	...	0.187
$m_1 = 0.320, m_2 = 0.065,$	x_2	0.001	0.003	0.005	0.007	0.009	0.011	0.013	0.015	...	0.065
$m_3 = 0.135, m_4 = 0.205,$	α_2^*	0.001	0.003	0.000	0.000	0.000	0.000	0.000	0.000	0.000	0.000
$m_5 = 0.275, t^{BR} = 0.033$	u_1	0.279	0.280	0.274	0.271	0.269	0.267	0.265	0.263	...	0.209

References

1. Alkalay-Houlihan, C., Shah, N.: The pure price of anarchy of pool block withholding attacks in bitcoin mining. In: AAAI, pp. 1724–1731. AAAI Press (2019)
2. Bag, S., Ruj, S., Sakurai, K.: Bitcoin block withholding attack: analysis and mitigation. IEEE Trans. Inf. Forensics Secur. **12**(8), 1967–1978 (2017)
3. Bag, S., Sakurai, K.: Yet another note on block withholding attack on bitcoin mining pools. In: Bishop, M., Nascimento, A.C.A. (eds.) ISC 2016. LNCS, vol. 9866, pp. 167–180. Springer, Cham (2016). https://doi.org/10.1007/978-3-319-45871-7_11
4. Bonneau, J.: Why buy when you can rent? Bribery attacks on bitcoin-style consensus. In: Clark, J., Meiklejohn, S., Ryan, P.Y.A., Wallach, D., Brenner, M., Rohloff, K. (eds.) FC 2016. LNCS, vol. 9604, pp. 19–26. Springer, Heidelberg (2016). https://doi.org/10.1007/978-3-662-53357-4_2
5. Chen, J., Micali, S.: Algorand: a secure and efficient distributed ledger. Theor. Comput. Sci. **777**, 155–183 (2019)
6. Chen, Z., Li, B., Shan, X., Sun, X., Zhang, J.: Discouraging pool block withholding attacks in bitcoins. CoRR abs/2008.06923 (2020)
7. Courtois, N.T., Bahack, L.: On subversive miner strategies and block withholding attack in bitcoin digital currency. CoRR abs/1402.1718 (2014)
8. Eyal, I.: The miner's dilemma. In: IEEE Symposium on Security and Privacy, pp. 89–103. IEEE Computer Society (2015)
9. Eyal, I., Sirer, E.G.: Majority is not enough: bitcoin mining is vulnerable. In: Christin, N., Safavi-Naini, R. (eds.) FC 2014. LNCS, vol. 8437, pp. 436–454. Springer, Heidelberg (2014). https://doi.org/10.1007/978-3-662-45472-5_28
10. Ke, J., Szalachowski, P., Zhou, J., Xu, Q., Yang, Z.: IBWH: an intermittent block withholding attack with optimal mining reward rate. In: Lin, Z., Papamanthou, C., Polychronakis, M. (eds.) ISC 2019. LNCS, vol. 11723, pp. 3–24. Springer, Cham (2019). https://doi.org/10.1007/978-3-030-30215-3_1
11. Li, W., Cao, M., Wang, Y., Tang, C., Lin, F.: Mining pool game model and nash equilibrium analysis for pow-based blockchain networks. IEEE Access **8**, 101049–101060 (2020)
12. Liao, K., Katz, J.: Incentivizing blockchain forks via whale transactions. In: Brenner, M., et al. (eds.) FC 2017. LNCS, vol. 10323, pp. 264–279. Springer, Cham (2017). https://doi.org/10.1007/978-3-319-70278-0_17

13. Luu, L., Saha, R., Parameshwaran, I., Saxena, P., Hobor, A.: On power splitting games in distributed computation: the case of bitcoin pooled mining. In: CSF, pp. 397–411. IEEE Computer Society (2015)
14. Mengelkamp, E., Nothei en, B., Beer, C., Dauer, D., Weinhardt, C.: A blockchain-based smart grid: towards sustainable local energy markets. Comput. Sci. Res. Dev. **33**(1-2), 207–214 (2018)
15. Nakamoto, S.: Bitcoin: a peer-to-peer electronic cash system (2008). https://bitcoin.org/bitcoin.pdf
16. Pinzón, C., Rocha, C.: Double-spend attack models with time advantage for bitcoin. In: CLEI Selected Papers. Electronic Notes in Theoretical Computer Science, vol. 329, pp. 79–103. Elsevier (2016)
17. Rosenfeld, M.: Analysis of bitcoin pooled mining reward systems. CoRR abs/1112.4980 (2011)
18. Rosenfeld, M.: Analysis of hashrate-based double spending. CoRR abs/1402.2009 (2014)
19. Sapirshtein, A., Sompolinsky, Y., Zohar, A.: Optimal selfish mining strategies in bitcoin. In: Grossklags, J., Preneel, B. (eds.) FC 2016. LNCS, vol. 9603, pp. 515–532. Springer, Heidelberg (2017). https://doi.org/10.1007/978-3-662-54970-4_30
20. Schrijvers, O., Bonneau, J., Boneh, D., Roughgarden, T.: Incentive compatibility of bitcoin mining pool reward functions. In: Grossklags, J., Preneel, B. (eds.) FC 2016. LNCS, vol. 9603, pp. 477–498. Springer, Heidelberg (2017). https://doi.org/10.1007/978-3-662-54970-4_28
21. Schwartz, D., Youngs, N., Britto, A., et al.: The ripple protocol consensus algorithm. Ripple Labs Inc White Paper **5**(8) (2014)
22. Wang, Q., Chen, Y.: The tight bound for pure price of anarchy in an extended miner's dilemma game. In: AAMAS, pp. 1695–1697. ACM (2021)
23. Wood, G., et al.: Ethereum: a secure decentralised generalised transaction ledger. Ethereum Project Yellow Paper **151**, 1–32 (2014)

Approximation Algorithms for the Directed Path Partition Problems

Yong Chen[1], Zhi-Zhong Chen[2], Curtis Kennedy[3], Guohui Lin[3(✉)], Yao Xu[4], and An Zhang[1]

[1] Department of Mathematics, Hangzhou Dianzi University, Hangzhou, China
{chenyong,anzhang}@hdu.edu.cn

[2] Division of Information System Design, Tokyo Denki University, Saitama, Japan
zzchen@mail.dendai.ac.jp

[3] Department of Computing Science, University of Alberta, Edmonton, Canada
{ckennedy,guohui}@ualberta.ca

[4] Department of Computer Science, Georgia Southern University, Statesboro, USA
yxu@georgiasouthern.edu

Abstract. Given a digraph $G = (V, E)$, the k-path partition problem is to find a minimum collection of vertex-disjoint directed paths each of order at most k to cover all the vertices of V. The problem has various applications in facility location, network monitoring, transportation and others. Its special case on undirected graphs has received much attention recently, but the general version is seemingly untouched in the literature. We present the first $k/2$-approximation algorithm, for any $k \geq 3$, based on a novel concept of augmenting path to minimize the number of singletons in the partition. When $k \geq 7$, we present an improved $(k + 2)/3$-approximation algorithm based on the maximum path-cycle cover followed by a careful 2-cycle elimination process. When $k = 3$, we define the second novel kind of augmenting paths to reduce the number of 2-paths and propose an improved 13/9-approximation algorithm.

Keywords: Path partition · digraph · augmenting path · matching · path-cycle cover · approximation algorithm

1 Introduction

Let G be a digraph. We denote the vertex set and the edge set of G by $V(G)$ and $E(G)$, respectively, and simplify them as V and E when G is clear from the context, that is, $G = (V, E)$. We assume without loss of generality that there are no self-loops or multiple edges in the graph. Let $n = |V|$ and $m = |E|$, which are referred to as the *order* and the *size* of the graph G, respectively. For a vertex v in G, the number of edges entering (leaving, respectively) v is denoted by $d^-_G(v)$ ($d^+_G(v)$, respectively), which is referred to as the *in-degree* (*out-degree*, respectively) of v. As well, $d^-_G(v)$ and $d^+_G(v)$ are simplified as $d^-(v)$ and $d^+(v)$, respectively, when the graph G is clear from the context.

J. Chen et al. (Eds.): IJTCS-FAW 2021, LNCS 12874, pp. 23–36, 2022.
https://doi.org/10.1007/978-3-030-97099-4_2

A *simple directed path* P in G is a subgraph in which the vertices can be ordered so that P contains exactly those edges from one vertex to the next vertex. Note that $|V(P)| = |E(P)| + 1$, that is, an order-k path is a length-$(k-1)$ path, and is simply called a *k-path*. For convenience, a 1-path is also called a *singleton*. Given a path, when there is an edge from the last vertex to the first vertex, then adding this edge to the path gives rise to a *simple directed cycle*. In the sequel, we leave out both "simple" and "directed", and simply call them a path and a cycle, respectively.

The *k-path partition* problem is to find a minimum collection of vertex-disjoint paths each of order at most k such that every vertex is on some path in the collection, for k being part of the input or a fixed integer. When k is a fixed integer, the problem is abbreviated as kPP, or otherwise as PP.

When the given graph is undirected, the edges are deemed bidirectional; this way, an undirected graph is a special digraph.

When $k \geq 3$, kPP on undirected graphs is NP-hard [7] and has received a number of studies [2–4,11,14]. Nevertheless, the general kPP is seemingly untouched in the literature. By ignoring edge directions, 2PP is equivalent to the Maximum Matching problem on undirected graphs, which is solvable in $O(m\sqrt{n}\log(n^2/m)/\log n)$-time [8]. Note that, since the Hamiltonian path problem is NP-hard [7], PP is APX-hard and is not approximable within a ratio of 2 unless P = NP.

One sees that kPP can be regarded as a special case of the minimum exact k-set cover, by creating a subset of ℓ vertices if and only if they form a directed path in the input graph, for all $\ell \leq k$. The minimum exact k-set cover problem is one of Karp's 21 NP-hard problems [9], and it does not admit any non-trivial approximation algorithms. The intractability of kPP and/or PP on special undirected graph classes (such as cographs, chordal, bipartite, comparability) has been investigated [10,12,13].

In the sequel, we assume $k \geq 3$ is a fixed constant and study kPP from the approximation algorithm perspective. kPP is solvable in polynomial time for trees [14], cographs [12] and bipartite permutation graphs [13]. For 3PP on undirected graphs, Monnot and Toulouse [11] presented a 3/2-approximation algorithm; the approximation ratio has been improved to 13/9 [3], 4/3 [2] and the currently best 21/16 [4]. For any fixed $k \geq 4$, kPP on undirected graphs admits a $k/2$-approximation algorithm [2].

It is noted that in various applications such as facility location, network monitoring, and transportation, the background network is modeled as a digraph. In this paper, we investigate the general kPP, and present the first $k/2$-approximation algorithm, for any $k \geq 3$, based on a novel concept of augmenting path to minimize the number of singletons in the partition. When $k \geq 7$, we present an improved $(k+2)/3$-approximation algorithm based on the maximum path-cycle cover followed by a careful 2-cycle elimination process. Lastly, for 3PP, we define the second novel kind of augmenting paths and propose an improved 13/9-approximation algorithm. The state-of-the-art approximation results are summarized in Table 1.

Table 1. The best known approximation ratios for kPP: the ones labeled with $*$ are achieved in this paper.

directed	$k \geq 3$	$k/2$-approx*
	$k \geq 7$	$(k+2)/3$-approx*
	$k = 3$	13/9-approx*
undirected	$k \geq 3$	$k/2$-approx [2] ($k = 3$ [11])
	$k \geq 7$	$(k+2)/3$-approx*
	$k = 3$	21/16-approx [4]

The rest of the paper is organized as follows: In Sect. 2 we present our three approximation algorithms, each in a separate subsection. For the last 13/9-approximation algorithm for 3PP, we also provide instances to show the tightness of the ratio. We conclude the paper in Sect. 3, with several future works.

2 Approximation Algorithms

Given a digraph $G = (V, E)$ and a positive integer b, a b-matching M in G is a spanning subgraph in which $d^+_M(v) \leq b$ and $d^-_M(v) \leq b$ for every vertex $v \in V$. One sees that a 1-matching in the digraph $G = (V, E)$ consists of vertex-disjoint paths and cycles, and thus it is also called a path-cycle cover. A maximum path-cycle cover of the graph G can be computed in $O(mn \log n)$ time [6].

We want to remind the readers that a b-matching can be defined in the same way for an undirected graph, where the edges are deemed bidirectional. Therefore, a 2-matching in an undirected graph consists of vertex-disjoint paths and cycles, which is also called a path-cycle cover. A 1-matching in an undirected graph is simply called a matching.

Below, for the purposes of computing a k-path partition in a digraph, we present a novel definition of *augmenting path*for decreasing the number of singletons in the partition iteratively. In fact, we are able to compute a k-path partition with the minimum number of singletons, leading to the first $k/2$-approximation for kPP. When $k = 3$, we present the second novel definition of augmenting path for decreasing the number of 2-paths in the partition iteratively. Though this time we are not able to achieve the minimum, we show that there are not too many 2-paths left compared to an optimal 3-path partition, leading to a 13/9-approximation for 3PP.

In Sect. 2.1, we present the first $k/2$-approximation for kPP for any $k \geq 3$. Section 2.2 deals with the case where $k \geq 7$; we start with a maximum path-cycle cover and then carefully deal with 2-cycles in the cover, leading to a $(k + 2)/3$-approximation algorithm. Lastly in Sect. 2.3, we design a 13/9-approximation algorithm for 3PP; we also show that the approximation ratio 13/9 of the last algorithm is tight.

2.1 The First $k/2$-Approximation for kPP

Suppose we are given a digraph $G = (V, E)$ and a k-path partition $\mathcal{Q}$ of G, where $\mathcal{Q}$ can be the collection of n singletons.

For an ℓ-path v_1-v_2-$\cdots$-v_ℓ in $\mathcal{Q}$, v_j is called the j-th vertex on the path, and in particular v_1 is the head vertex and v_ℓ is the tail vertex. The intention of a to-be-defined augmenting path is to reduce the number of singletons, by adding an edge, so as to improve $\mathcal{Q}$.

Let us first use $\mathcal{Q}$ to define two types of edges that can be on our desired alternating paths. We require that each alternating path starts with a singleton. If there is no singleton in $\mathcal{Q}$, then we do not bother to define the edge types and $\mathcal{Q}$ is our *desired solution* achieving the minimum number of singletons. In the other case, for every ℓ-path v_1-v_2-$\cdots$-v_ℓ in $\mathcal{Q}$, $\ell = 2, 3, \ldots k$, the edges (v_1, v_2) and $(v_{\ell-1}, v_\ell)$ are *matched* edges; all the edges of $E(G) - E(\mathcal{Q})$ incident at v_1 (v_ℓ, respectively), both entering and leaving v_1 (v_ℓ, respectively), are *free* edges with respect to $\mathcal{Q}$. We note that there are edges which are neither matched nor free, such as the edge (v_2, v_3) when $\ell \geq 4$; these are *irrelevant* edges never being part of alternating paths.

We next explain how to grow an alternating path. Consider the last free edge (u, v) on the alternating path extending to v. (The following argument applies to the symmetric case (v, u) by reversing the direction of the involved edges, if any.) If there is no matched edge entering v, that is, v is not the second vertex of a path of $\mathcal{Q}$ (see Fig. 1a for illustrations), then the alternating path ends. In Lemma 2, we will show that such an alternating path is an augmenting path, and it can be used to transfer $\mathcal{Q}$ into another k-path partition (by taking the *symmetric difference* of $\mathcal{Q}$ and the edges on the augmenting path) with at least one less singleton.

If there is a matched edge entering v, that is, v is the second vertex v_2 of an ℓ-path v_1-v_2-$\cdots$-v_ℓ of $\mathcal{Q}$, but the matched edge (v_1, v_2) has already been included in the alternating path, then the alternating path ends too and it is not an augmenting path (but an alternating cycle).

In the other case where v is the second vertex v_2 of an ℓ-path v_1-v_2-$\cdots$-v_ℓ of $\mathcal{Q}$ ($\ell \leq k$; see Fig. 1b for an illustration), the matched edge (v_1, v_2), which is shown as the dotted edge in Fig. 1b, extends the alternating to the vertex v_1. Then, iteratively, when there is no free edge incident at v_1, the alternating path ends resulting in no augmenting path; or otherwise a free edge extends the alternating path and we may repeat the above process on this newly added free edge. Therefore, either a free edge or a matched edge ends the alternating path, and an augmenting path is achieved if and only if a free edge ends the alternating path so that there is no matched edge incident at the last vertex. In Lemma 1, we show that the entire process can be done via a breadth-first-search (BFS) traversal and takes time $O(m)$.

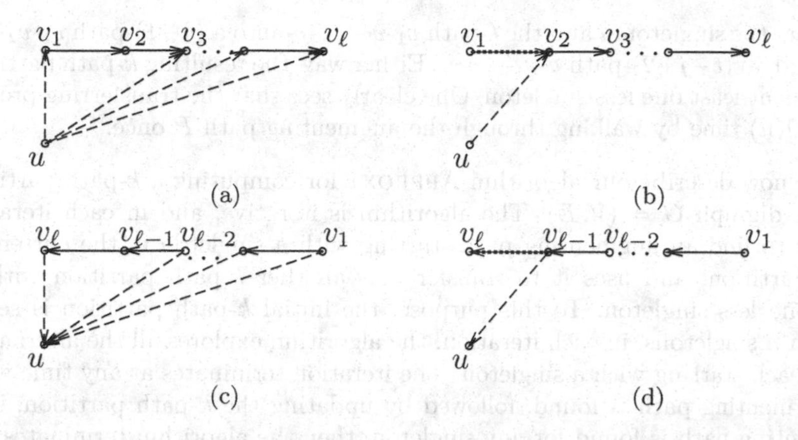

Fig. 1. (a) and (b), a free edge (u, v) for all possible configurations of the vertex v in the k-path partition $\mathcal{Q}$, where the dashed edges are free, the solid edges are in $\mathcal{Q}$, and the additionally dotted edge is matched. (c) and (d) show the symmetric case for a free edge (v, u). An alternating path is one alternating free and matched edges, and it starts with a singleton.

Lemma 1. *Given a k-path partition $\mathcal{Q}$ in the digraph $G = (V, E)$, determining whether or not there exists an augmenting path with respect to $\mathcal{Q}$, and if so finding one such path, can be done in $O(m)$ time.*

Proof. A single BFS traversal is sufficient since when two alternating paths starting with two distinct singletons meet, they share everything from that point on and no re-exploration is necessary. □

Lemma 2. *Given a k-path partition $\mathcal{Q}$ in the digraph $G = (V, E)$, if there exists an augmenting path with respect to $\mathcal{Q}$, then $\mathcal{Q}$ can be transferred into another k-path partition with at least one less singleton in $O(n)$ time.*

Proof. Let P denote the augmenting path with respect to $\mathcal{Q}$, and s denote its starting vertex which is a singleton. Let (u, v) denote the free edge which ends P.

The prefix of P from s to u, denoted as $P(s, u)$, is an even length alternating path. Replacing the matched edges of $\mathcal{Q}$ on $P(s, u)$ by the free edges on $P(s, u)$ transfers $\mathcal{Q}$ into another k-path partition $\mathcal{Q}'$ with exactly the same number of i-paths, for each $i = 1, 2, \ldots, k$, and additionally, in $\mathcal{Q}'$ the vertex u becomes a singleton and, without loss of generality, v is the j-th vertex v_j, where $j \neq 2$, on an ℓ-path v_1-v_2-$\cdots$-v_ℓ of $\mathcal{Q}'$, see Fig. 1a for illustrations.

It follows that, when $\ell \leq 2$, $j = 1$ and adding the free edge (u, v_1) to $\mathcal{Q}'$ merges u and the ℓ-path v_1-v_2-$\cdots$-v_ℓ into an $(\ell + 1)$-path (see Fig. 1a for an illustration). When $\ell \geq 3$ and $j = 1$, adding the free edge (u, v_1) to while removing the edge (v_1, v_2) from $\mathcal{Q}'$ transfers the singleton u and the ℓ-path v_1-v_2-$\cdots$-v_ℓ into a 2-path u-v_1 and an $(\ell - 1)$-path v_2-$\cdots$-v_ℓ. When $\ell \geq 3$ and $j \geq 3$, adding the free edge (u, v_j) to while removing the edge (v_{j-1}, v_j) from $\mathcal{Q}'$

transfers the singleton u and the ℓ-path v_1-v_2-$\cdots$-v_ℓ into a $(j-1)$-path v_1-v_2-$\cdots$-v_{j-1} and an $(\ell-j+2)$-path u-v_j-$\cdots$-v_ℓ. Either way, the resulting k-path partition contains at least one less singleton. One clearly sees that the transferring process takes $O(n)$ time by walking through the augmenting path P once. □

We now describe our algorithm APPROX1 for computing a k-path partition $\mathcal{Q}$ in a digraph $G = (V, E)$. The algorithm is iterative, and in each iteration it tries to find an augmenting path starting with a singleton in the current k-path partition, and uses it to transfer into another k-path partition with at least one less singleton. To this purpose, the initial k-path partition is set to contain n singletons; in each iteration, the algorithm explores all the alternating paths each starting with a singleton. The iteration terminates at any time when an augmenting path is found, followed by updating the k-path partition; if no augmenting path is found for any singleton, then the algorithm terminates and returns the current k-path partition as the solution. From Lemmas 1 and 2, the overall running time is $O(nm)$. A high level description of the algorithm is depicted in Fig. 2.

Algorithm APPROX1:
Input: a digraph $G = (V, E)$;
Output: a k-path partition $\mathcal{Q}$.

1. $\mathcal{Q}$ is initialized to contain n singletons;
2. For each singleton u in $\mathcal{Q}$,
   2.1 explore the alternating paths starting with u via a BFS traversal;
   2.2 if an augmenting path is found, update $\mathcal{Q}$ and break to restart Step 2;
3. Return $\mathcal{Q}$.

Fig. 2. A high level description of the algorithm APPROX1.

Theorem 1 [1]. *The algorithm* APPROX1 *is an $O(nm)$-time $k/2$-approximation for kPP.*

Proof. We first show that the k-path partition $\mathcal{Q}$ returned by the algorithm APPROX1 achieves the minimum number of singletons among all k-path partitions. The proof is done by constructing a mapping from the singletons in $\mathcal{Q}$ to the singletons of any other k-path partition $\mathcal{Q}'$, using the alternating paths with respect to $\mathcal{Q}$. First, if s is a singleton in both $\mathcal{Q}$ and $\mathcal{Q}'$, then s is mapped to s. Below we consider s being a singleton in $\mathcal{Q}$ but not in $\mathcal{Q}'$. We assume w.l.o.g. that the edge of $\mathcal{Q}'$ incident at s leaves s, that is, (s, v). So (s, v) is a free edge with respect to $\mathcal{Q}$.

Due to the non-existence of an augmenting path, we conclude that v is the second vertex v_2 of an ℓ-path v_1-v_2-$\cdots$-v_ℓ of $\mathcal{Q}$ (see Fig. 1b for an illustration). Since the free edge (s, v_2) is in $\mathcal{Q}'$, the matched edge (v_1, v_2) cannot be in $\mathcal{Q}'$ and it can be "discovered" only by the free edge (s, v_2) of $\mathcal{Q}'$.

We distinguish two cases: In Case 1, v_1 is a singleton in $\mathcal{Q}'$. Then, adding the free edge (s, v_2) to and removing the matched edge (v_1, v_2) from $\mathcal{Q}$ will transfer $\mathcal{Q}$ into another k-path partition in which s is no longer a singleton but v_1 becomes a singleton. In this sense, we say that the alternating path *maps* the singleton s of $\mathcal{Q}$ to the singleton v_1 of $\mathcal{Q}'$.

In Case 2, v_1 is not a singleton in $\mathcal{Q}'$. The edge of $\mathcal{Q}'$ incident at v_1, either entering or leaving v_1, is a free edge with respect to $\mathcal{Q}$. We assume w.l.o.g. that this edge leaves v_1, that is, (v_1, w). One sees that v_1 takes up the same role as the singleton s in the above argument, and again by the algorithm w has to be the second vertex w_2 of some ℓ'-path w_1-w_2-$\cdots$-$w_{\ell'}$ of $\mathcal{Q}$. Note that similarly the matched edge (w_1, w_2) cannot be in $\mathcal{Q}'$ and it can be "discovered" only by the free edge (v_1, w_2) of $\mathcal{Q}'$. We then repeat the above discussion on w_1, either to have an alternating path mapping the singleton s of $\mathcal{Q}$ to the singleton w_1 of $\mathcal{Q}'$, or to use the free edge in $\mathcal{Q}'$ and incident at w_1 to extend the alternating path. (*Symmetrically, if the free edge of* $\mathcal{Q}'$ *enters* v_1, *that is,* (w, v_1), *then* w *has to be the second last vertex* $w_{\ell'-1}$ *of some* ℓ'*-path* w_1-w_2-$\cdots$-$w_{\ell'}$ *of* $\mathcal{Q}$. *Similarly the matched edge* $(w_{\ell'-1}, w_{\ell'})$ *cannot be in* $\mathcal{Q}'$ *and it can be "discovered" only by the free edge* $(w_{\ell'-1}, v_1)$ *of* $\mathcal{Q}'$. *Consequently, we either have an alternating path mapping the singleton* s *of* $\mathcal{Q}$ *to the singleton* $w_{\ell'}$ *of* $\mathcal{Q}'$, *or use the free edge in* $\mathcal{Q}'$ *and incident at* $w_{\ell'}$ *to extend the alternating path.* See Fig. 1d for an illustration.) Due to the finite order of the graph G, at the end we will have an alternating path mapping the singleton s of $\mathcal{Q}$ to a singleton of $\mathcal{Q}'$.

Using the fact that there is at most one edge of $\mathcal{Q}$ ($\mathcal{Q}'$, respectively) leaving each vertex and at most one edge of $\mathcal{Q}$ ($\mathcal{Q}'$, respectively) entering each vertex, a singleton of $\mathcal{Q}'$ is not mapped by more than one singleton of $\mathcal{Q}$. Since every singleton of $\mathcal{Q}$ is mapped to a singleton of $\mathcal{Q}'$, we conclude that the number of singletons in $\mathcal{Q}$ is no greater than the number of singletons in $\mathcal{Q}'$.

Let $\mathcal{Q}^*$ denote an optimal k-path partition that minimizes the number of paths. Also, let $\mathcal{Q}_i^*$ ($\mathcal{Q}_i$, respectively) denote the sub-collection of all the i-paths of $\mathcal{Q}^*$ ($\mathcal{Q}$, respectively), for $i = 1, 2, \ldots, k$. It follows that

$$\sum_{i=1}^{k} i|\mathcal{Q}_i| = n = \sum_{i=1}^{k} i|\mathcal{Q}_i^*| \text{ and } |\mathcal{Q}_1| \le |\mathcal{Q}_1^*|.$$

Adding them together we have

$$2|\mathcal{Q}_1| + 2|\mathcal{Q}_2| + 3|\mathcal{Q}_3| + \cdots + k|\mathcal{Q}_k| \le 2|\mathcal{Q}_1^*| + 2|\mathcal{Q}_2^*| + 3|\mathcal{Q}_3^*| + \cdots + k|\mathcal{Q}_k^*|,$$

which leads to $2|\mathcal{Q}| \le k|\mathcal{Q}^*|$ and thus proves the theorem. □

2.2 An Improved $(k+2)/3$-Approximation for kPP, When $k \ge 7$

Throughout this section, we fix a digraph G and an integer $k \ge 7$. Our $(k+2)/3$-approximation algorithm for kPP starts by performing the following three steps:

Algorithm Approx2:

1. Compute a maximum path-cycle cover $\mathcal{C}$ of G.
2. While there is an edge $(u, v) \in E(G) - E(\mathcal{C})$ such that $d^+_{\mathcal{C}}(u) = 0$ (respectively, $d^-_{\mathcal{C}}(v) = 0$) and v (respectively, u) is a vertex of some cycle C of $\mathcal{C}$, modify $\mathcal{C}$ by deleting the edge entering v (respectively, leaving u) in $\mathcal{C}$ and adding edge (u, v).
3. Construct a digraph $G_1 = (V(G), E_1)$, where E_1 is the set of all edges $(u, v) \in E(G) - E(\mathcal{C})$ such that u and v appear in different connected components of $\mathcal{C}$ and at least one of u and v appears on a 2-cycle of $\mathcal{C}$.

Hereafter, $\mathcal{C}$ always refers to the path-cycle cover obtained after the completion of Step 2. We give several definitions related to the graphs G_1 and $\mathcal{C}$. A *path* (respectively, *cycle*) *component* of $\mathcal{C}$ is a connected component of $\mathcal{C}$ that is a path (respectively, cycle). Let S be a subgraph of G_1. S *saturates* a 2-cycle C of $\mathcal{C}$ if at least one edge of S is incident at a vertex of C. The *weight* of S is the number of 2-cycles of $\mathcal{C}$ saturated by S. For convenience, we say that two connected components C_1 and C_2 of $\mathcal{C}$ are *adjacent* in a subgraph G' of G if there is an edge $(u_1, u_2) \in E(G')$ such that $u_1 \in V(C_1)$ and $u_2 \in V(C_2)$.

We are able to prove that a maximum-weight path-cycle cover in G_1 can be computed in $O(nm \log n)$ time. Due to space limit, the proof is supplied in the full version of the paper [1]. Our algorithm then proceeds to perform the following four steps:

4. Compute a maximum-weight path-cycle cover M in G_1.
5. While there is an edge $e \in E(M)$ such that the removal of e from M does not change the weight of M, delete e from M.
6. Construct a digraph $G_2 = (V(G), E(\mathcal{C}) \cup E(M))$. (*Comment:* For each pair of connected components of $\mathcal{C}$, there is at most one edge between them in G_2 because of Step 5.)
7. Construct an undirected graph G_3, where the vertices of G_3 one-to-one correspond to the connected components of $\mathcal{C}$ and two vertices are adjacent in G_3 if and only if the corresponding connected components of $\mathcal{C}$ are adjacent in G_2.

An undirected graph G' is a *star* if G' is a connected graph with at least one edge and all but at most one vertex of G' are of degree 1 in G'. If a star G' has a vertex of degree larger than 1, then this unique vertex is the *center* of G'; otherwise, G' is an edge and we choose an arbitrary vertex of G' as the *center* of G'. Each vertex of a star G' other than its center is a *satellite* of G'. A vertex u is *isolated* in an undirected graph G' if its degree in G' is 0.

Now, we are ready to state the final steps of our algorithm, where the detailed transformation algorithms in Steps 9–10, as well as Lemmas 3–5, can be found in [1].

8. For each connected component K of G_3 that is a vertex corresponding to a 2-cycle C in $\mathcal{C}$, transform C into a k-path partition of $G[V(C)]$ by deleting an arbitrary edge of C.
9. For each connected component K of G_3 that is a vertex corresponding to no 2-cycle in $\mathcal{C}$, transform $G_2[V(C)]$ into a k-path partition $\mathcal{P}$ of $G[V(C)]$ such that $|E(\mathcal{P})| \geq \frac{2}{3}|E(C)|$ [1, Lemma 3].
10. For each connected component K of G_3 that is a star whose center corresponds to a path (respectively, cycle) component of $\mathcal{C}$, transform $G_2[U]$ into a k-path partition $\mathcal{P}$ of $G[U]$ such that $|E(\mathcal{P})| \geq \frac{2}{3}|F|$, where $U = \bigcup_{C \in K} V(C)$ and $F = \bigcup_{C \in K} E(C)$ [1, Lemmas 4 and 5, respectively].
11. Output the graph $\mathcal{Q} = (V(G), \bigcup_{\mathcal{P}} E(\mathcal{P}))$ as a k-path partition of G, where $\mathcal{P}$ ranges over all k-path partitions obtained in Steps 8–10.

The time complexity of our algorithm is dominated by the computation of $\mathcal{C}$ and M. So, the overall time complexity is $O(nm \log n)$. Finally, we analyze the approximation ratio achieved by our algorithm. To this end, let $\mathcal{Q}$ be the k-path partition outputted by our algorithm; let $\mathcal{Q}^*$ be an optimal solution of G for kPP. Obviously, $|E(\mathcal{C})| \geq |E(\mathcal{Q}^*)|$.

An *isolated 2-cycle* of G_2 is a 2-cycle of G_2 whose corresponding vertex in G_3 is isolated in G_3. Let $\mathcal{I}$ be the set of isolated 2-cycles in G_2. We have:

- $|E(\mathcal{Q})| \geq |\mathcal{I}| + \frac{2}{3}(|E(\mathcal{C})| - 2|\mathcal{I}|) = \frac{2}{3}|E(\mathcal{C})| - \frac{1}{3}|\mathcal{I}| \geq \frac{2}{3}|E(\mathcal{Q}^*)| - \frac{1}{3}|\mathcal{I}|$ (by Steps 8–10, [1, Lemmas 3–5]);
- $|E(\mathcal{Q}^*)| \leq \frac{k-1}{k} \cdot (n - 2|\mathcal{I}|) + |\mathcal{I}| = \frac{k-1}{k}n - \frac{k-2}{k}|\mathcal{I}|$ [1, Lemma 2];
- $|\mathcal{Q}| = n - |E(\mathcal{Q})|$ and $|\mathcal{Q}^*| = n - |E(\mathcal{Q}^*)|$.

Therefore, the approximation ratio achieved by our algorithm is $\frac{|\mathcal{Q}|}{|\mathcal{Q}^*|} = \frac{n-|E(\mathcal{Q})|}{n-|E(\mathcal{Q}^*)|} \leq \frac{n-\frac{2}{3}|E(\mathcal{Q}^*)|+\frac{1}{3}|\mathcal{I}|}{n-|E(\mathcal{Q}^*)|}$. By calculus, one can verify that the last fraction is an increasing function in $|E(\mathcal{Q}^*)|$. So,

$$\frac{|\mathcal{Q}|}{|\mathcal{Q}^*|} \leq \frac{n - \frac{2}{3}\left(\frac{k-1}{k}n - \frac{k-2}{k}|\mathcal{I}|\right) + \frac{1}{3}|\mathcal{I}|}{n - \left(\frac{k-1}{k}n - \frac{k-2}{k}|\mathcal{I}|\right)} = \frac{\frac{k+2}{3k}n + \frac{3k-4}{3k}|\mathcal{I}|}{\frac{1}{k}n + \frac{k-2}{k}|\mathcal{I}|} \leq \max\left\{\frac{k+2}{3}, \frac{3k-4}{3k-6}\right\}.$$

Now, since $k \geq 7$, $\frac{k+2}{3} \geq \frac{3k-4}{3k-6}$. Therefore, $\frac{|\mathcal{Q}|}{|\mathcal{Q}^*|} \leq \frac{k+2}{3}$.

Summarizing up, we have the following theorem.

Theorem 2. *The algorithm* APPROX2 *is an* $O(nm \log n)$*-time* $(k+2)/3$*-approximation for* k*PP, where* $k \geq 7$.

2.3 A 13/9-Approximation for 3PP

In this section, we present another set of definitions of matched edges and free edges, alternating paths, and augmenting paths to reduce the number of 2-paths in a 3-path partition $\mathcal{Q}$.

This time, the edges on the 2-paths of $\mathcal{Q}$ are *matched* edges and the edges outside of $\mathcal{Q}$ are *free edges.* (The edges on the 3-paths of $\mathcal{Q}$ are *irrelevant.*) We require that an augmenting path starts with a matched edge then a free edge, which form a 3-path, alternating matched and free edges, and lastly ends with a matched edge. The intention is to convert three 2-paths of $\mathcal{Q}$ into two 3-paths, and thus the augmenting path should contain at least three distinct matched edges. Formally, the following constraints must be satisfied:

1) An augmenting path starts with a matched edge, alternating free and matched edges, ends with a matched edge, and contains at least three distinct matched edges;
2) only the first and the last matched edges can be included twice (Fig. 3b), and if they are the same edge, then this edge is included exactly three times (Fig. 3c);
3) the first matched edge and the first free edge form a 3-path in the graph G (Fig. 3a);
4) if the first matched edge is included twice, then the first free edge and its adjacent free edge on the augmenting path form a 3-path (Fig. 3b);
5) if the last matched edge is not included twice, then it and the last free edge form a 3-path (Fig. 3a);
6) if the last matched edge is included twice, then the last free edge and its adjacent free edge on the augmenting path form a 3-path (Fig. 3b).

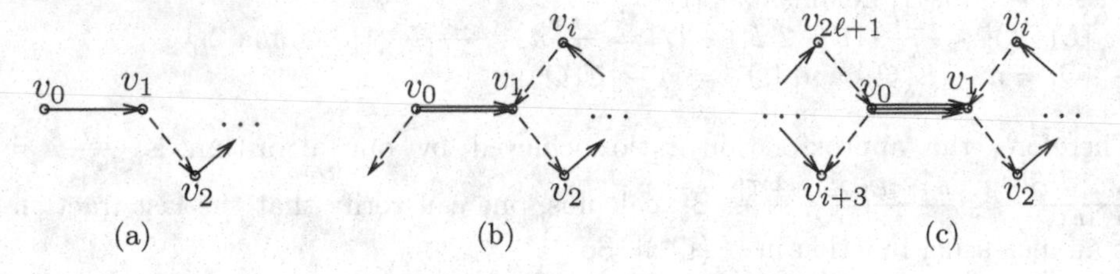

Fig. 3. Local configurations of an augmenting path: (a) The first matched edge (v_0, v_1) and the first free edge (v_1, v_2) form a 3-path v_0-v_1-v_2 in the graph G; symmetrically, if the last matched edge is not included twice, then it and the last free edge form a 3-path. (b) The first matched edge (v_0, v_1) is included twice; the first free edge (v_1, v_2) and its adjacent free edge (v_i, v_1) on the augmenting path form a 3-path v_i-v_1-v_2. (c) The first and the last matched edges are the same edge (v_0, v_1); the first free edge (v_1, v_2) and its adjacent free edge (v_i, v_1) on the augmenting path form a 3-path v_i-v_1-v_2, and the last free edge $(v_{2\ell+1}, v_0)$ and its adjacent free edge (v_0, v_{i+3}) on the augmenting path form a 3-path $v_{2\ell+1}$-v_0-v_{i+3}.

Upon an augmenting path, an analogous symmetric difference adds its free edges to while removes its internal matched edges from $\mathcal{Q}$ (that is, if the first/last matched edge on the augmenting path appears only once, then it is kept). This way, the collection of ℓ 2-paths of $\mathcal{Q}$ is transferred into $(\ell - 3)$ 2-paths and two 3-paths on the same set of vertices, here ℓ denotes the number of distinct matched edges on the augmenting path. We point out that, if the first/last matched edge is included twice, then it is removed during the symmetric difference since one copy is internal on the augmenting path. One sees that the net effect is to transfer three 2-paths into two 3-paths, thus reducing the number of paths by 1. Also, during such processes, no singletons or existing 3-paths of $\mathcal{Q}$ are touched. We initialize the 3-path partition $\mathcal{Q}$ to be the solution produced by the algorithm Approx1.

To find an augmenting path, we define below the alternating paths and explain how to grow them. Each alternating path starts with a matched edge and then a free edge such that these two edges form a 3-path in the graph G (i.e., satisfying Constraint #3). Consider w.l.o.g. a matched edge (v_0, v_1) followed by a free edge (v_1, v_2), see Fig. 3 for an illustration. If the vertex v_2 is not incident with a matched edge, then the alternating path is not extendable; or otherwise the unique matched edge incident at v_2 extends the alternating path. Note that there is no direction requirement on the second matched edge and we w.l.o.g. assume it is (v_2, v_3). Next, similarly, if the vertex v_3 is not incident with any free edge, then the alternating path is not extendable; or otherwise a free edge incident at v_3 extends the alternating path and the extending process goes on. We remark that v_2 can collide into v_0, and if so (i.e., (v_1, v_0) is a free edge), then the edge (v_0, v_1) is included the second time into the alternating path and the two free edges incident at the vertex v_1 (either added already or to be added next) must form a 3-path in the graph G, that is, Constraint #4 must be satisfied for the extending process to go on. Indeed, during the extending process, whenever a matched edge e is included, Constraint #2 is checked and there are three possible cases:

Case 1. e appears the first time on the alternating path. Then Constraints #1, #5 are checked:
- **1.1.** If both are satisfied, then an augmenting path is achieved;
- **1.2.** otherwise the extending process goes on.

Case 2. e appears the second time on the alternating path.
- **2.1.** If e is the same as the first matched edge and Constraint #4 is satisfied or can be satisfied, then the extending process goes on (and use a free edge to satisfy Constraint #4, if necessary);
- **2.2.** if e is not the same as the first matched edge and Constraint #6 is satisfied, then an augmenting path is achieved;
- **2.3.** otherwise the extending process terminates.

Case 3. e appears the third time on the alternating path. Then Constraints #4, #6 are checked:
- **3.1.** If both are satisfied, then an augmenting path is achieved;
- **3.2.** otherwise the extending process terminates.

In the above, by "the extending process goes on" we mean to use a free edge incident at the last vertex to extend the alternating path, with an additional consideration in Case 2.1 where such an edge might have to satisfy Constraint #4. And then the matched edge incident at the last vertex extends the alternating path. At the non-existence of such a free edge or such a matched edge, the alternating path is not extendable. By "the extending process terminates", we mean that the alternating path does not lead to an augmenting path and thus the process is early terminated.

In [1], we show that given a digraph $G = (V, E)$ and a 3-path partition $\mathcal{Q}$ with the minimum number of singletons, determining whether or not there exists an augmenting path with respect to $\mathcal{Q}$, and if so finding one such path, can be done through $O(n)$ BFS traversals and thus in $O(nm)$ time. Furthermore, if an augmenting path with respect to $\mathcal{Q}$ is identified, then $\mathcal{Q}$ can be transferred in $O(n)$ time into another 3-path partition with the same number of singletons, three less 2-paths, and two more 3-paths (that is, a net reduction of one path).

Our algorithm Approx3 starts with the 3-path partition $\mathcal{Q}$ returned by the algorithm Approx1 for the input digraph $G = (V, E)$. Recall that $\mathcal{Q}$ contains the minimum number of singletons. The same as Approx1, the algorithm Approx3 is iterative too, and in each iteration it tries to find an augmenting path starting with a matched edge with respect to the current 3-path partition $\mathcal{Q}$, and uses it to update $\mathcal{Q}$ to have the same number of singletons, three less 2-paths, and two more 3-paths. To this purpose, the algorithm explores all the alternating paths each starting with a matched edge, and the iteration terminates at any time when an augmenting path is found, followed by updating the 3-path partition. If in an iteration no augmenting path is found for any matched edge, then the algorithm Approx3 terminates and returns the achieved 3-path partition as the solution. A high level description of the algorithm is depicted in Fig. 4, with its overall running time in $O(n^2 m)$.

Algorithm Approx3:
Input: a digraph $G = (V, E)$;
Output: a 3-path partition $\mathcal{Q}$.

1. $\mathcal{Q}$ is initialized to be the 3-path partition for G computed by Approx1.
2. For each matched edge e in $\mathcal{Q}$,
   2.1 explore the alternating paths starting with e via a BFS traversal;
   2.2 if an augmenting path is found, update $\mathcal{Q}$ and break to restart Step 2;
3. Return $\mathcal{Q}$.

Fig. 4. A high level description of the algorithm Approx3.

Theorem 3. *The algorithm* Approx3 *is an* $O(n^2 m)$*-time* 13/9*-approximation for* 3PP.

Proof. Let $\mathcal{Q}^*$ denote an optimal 3-path partition that minimizes the number of paths. Also, let $\mathcal{Q}_i^*$ ($\mathcal{Q}_i$, respectively) denote the sub-collection of all the i-paths of $\mathcal{Q}^*$ ($\mathcal{Q}$, respectively), for $i = 1, 2, 3$. It follows that

$$|\mathcal{Q}_1| + 2|\mathcal{Q}_2| + 3|\mathcal{Q}_3| = n = |\mathcal{Q}_1^*| + 2|\mathcal{Q}_2^*| + 3|\mathcal{Q}_3^*| \text{ and } |\mathcal{Q}_1| \le |\mathcal{Q}_1^*|.$$

We develop a counting argument to prove that $|\mathcal{Q}_2| \le |\mathcal{Q}_1^*| + 2|\mathcal{Q}_2^*| + 4|\mathcal{Q}_3^*|/3$, that is, there are not too many 2-paths left in $\mathcal{Q}$ compared to $\mathcal{Q}^*$, and thus prove the performance ratio of 13/9.

Due to space limit, the complete proof is supplied in [1]. □

A tight instance for Approx3. The tight instance for the 13/9-approximation algorithm for 3PP on undirected graphs in [3] can be modified to show the tightness of Approx3, as illustrated in Fig. 5. One sees that for this digraph, $|\mathcal{Q}_{3,3}^*| = 6 = 2|\mathcal{Q}_3^*|/3$, suggesting the performance analysis for the algorithm Approx3 is tight.

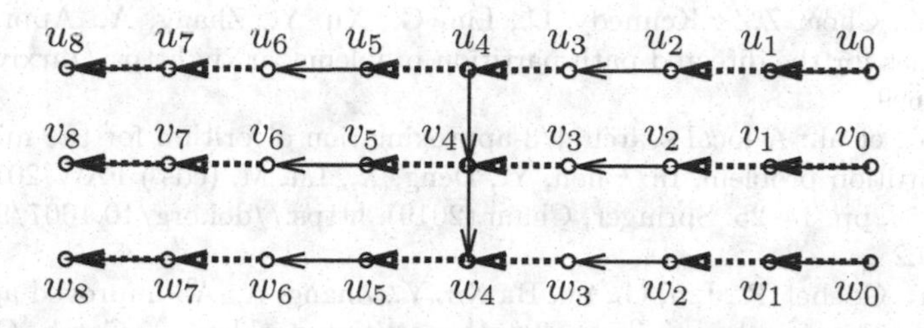

Fig. 5. A tight instance of 27 vertices, in which the 3-path partition $\mathcal{Q}$ produced by the algorithm Approx3 contains twelve 2-paths and one 3-path (solid edges) and an optimal 3-path partition $\mathcal{Q}^*$ contains nine 3-paths (dashed edges). The six edges (u_0, u_1), (u_7, u_8), (v_0, v_1), (v_7, v_8), (w_0, w_1), (w_7, w_8) are in $\mathcal{Q}_2$ and in $\mathcal{Q}_3^*$, shown in both solid and dashed (they are drawn overlapping). Note that the two edges (u_4, v_4), (v_4, w_4) of $\mathcal{Q}_3$ are irrelevant edges, and there is no augmenting path with respect to $\mathcal{Q}$.

3 Final Remarks

We studied the general kPP problem on digraphs, which seemingly escaped from the literature. We proposed a novel concept of augmenting path to design the first $k/2$-approximation algorithm for the problem, which is iterative and in each iteration it seeks to reduce the number of singletons until impossible. When $k \ge 7$, we were able to design an improved $(k+2)/3$-approximation algorithm, starting with the maximum path-cycle cover in the graph to carefully eliminate the 2-cycles. Certainly, this is also a $(k+2)/3$-approximation algorithm for the special case of undirected graphs, improving the previously best approximation ratio of $k/2$ [2].

When $k = 3$, we defined the second kind of augmenting paths to reduce the number of 2-paths and presented an improved 13/9-approximation algorithm.

See Table 1 for the summarized approximation results as of today. Designing better approximation algorithms for kPP, in any listed case, is certainly interesting, in particular, for the general 3PP. Above all, an $o(k)$-approximation algorithm for kPP would be exciting.

On the other hand, when k is part of the input, the k-path partition problem is APX-hard and can not be approximated within a ratio of 2. It would be interesting to know whether kPP is APX-hard or not, for any fixed $k \geq 3$, and to see some non-trivial lower bounds on the approximation ratios.

Acknowledgments. This research is supported by the NSFC Grants 11771114 and 11971139 (YC and AZ), the Zhejiang Provincial NSFC Grant LY21A010014 (YC and AZ), the CSC Grants 201508330054 (YC) and 201908330090 (AZ), the Grant-in-Aid for Scientific Research of the Ministry of Education, Science, Sports and Culture of Japan Grant No. 18K11183 (ZZC), and the NSERC Canada (GL).

References

1. Chen, Y., Chen, Z.Z., Kennedy, C., Lin, G., Xu, Y., Zhang, A.: Approximation algorithms for the directed path partition problems. arXiv https://arxiv.org/abs/2107.04699
2. Chen, Y., et al.: A local search 4/3-approximation algorithm for the minimum 3-path partition problem. In: Chen, Y., Deng, X., Lu, M. (eds.) FAW 2019. LNCS, vol. 11458, pp. 14–25. Springer, Cham (2019). https://doi.org/10.1007/978-3-030-18126-0_2
3. Chen, Y., Goebel, R., Lin, G., Su, B., Xu, Y., Zhang, A.: An improved approximation algorithm for the minimum 3-path partition problem. J. Comb. Optim. **38**, 150–164 (2019). https://doi.org/10.1007/s10878-018-00372-z
4. Chen, Y., Goebel, R., Su, B., Tong, W., Xu, Y., Zhang, A.: A 21/16-approximation for the minimum 3-path partition problem. In: Proceedings of ISAAC 2019. LIPIcs, vol. 149, pp. 46:1–46:20 (2019)
5. Cormen, T.H., Leiserson, C.E., Rivest, R.L., Stein, C.: Introduction to Algorithms, 3rd edn. The MIT Press, Cambridge (2009)
6. Gabow, H.N.: An efficient reduction technique for degree-constrained subgraph and bidirected network flow problems. In: Proceedings of the 15th Annual ACM Symposium on Theory of Computing (STOC 1983), pp. 448–456 (1983)
7. Garey, M.R., Johnson, D.S.: Computers and Intractability: A Guide to the Theory of NP-Completeness. W. H. Freeman and Company, San Francisco (1979)
8. Goldberg, A.V., Karzanov, A.V.: Maximum skew-symmetric flows and matchings. Math. Program. **100**, 537–568 (2004). https://doi.org/10.1007/s10107-004-0505-z
9. Karp, R.M.: Reducibility among combinatorial problems. In: Miller, R.E., Thatcher, J.W., Bohlinger, J.D. (eds.) Complexity of Computer Computations. IRSS, pp. 85–103. Springer, Boston (1972). https://doi.org/10.1007/978-1-4684-2001-2_9
10. Korpelainen, N.: A boundary class for the k-path partition problem. Electron. Notes Discrete Math. **67**, 49–56 (2018)
11. Monnot, J., Toulouse, S.: The path partition problem and related problems in bipartite graphs. Oper. Res. Lett. **35**, 677–684 (2007)
12. Steiner, G.: On the k-th path partition problem in cographs. Congr. Numer. **147**, 89–96 (2000)
13. Steiner, G.: On the k-path partition of graphs. Theor. Comput. Sci. **290**, 2147–2155 (2003)
14. Yan, J.H., Chang, G.J., Hedetniemi, S.M., Hedetniemi, S.T.: k-path partitions in trees. Discrete Appl. Math. **78**, 227–233 (1997)

Faster Algorithms for k-Subset Sum and Variations

Antonis Antonopoulos, Aris Pagourtzis(✉), Stavros Petsalakis, and Manolis Vasilakis

School of Electrical and Computer Engineering, National Technical University of Athens, Polytechnioupoli, 15780 Zografou, Greece
{aanton,spetsalakis,mvasilakis}@corelab.ntua.gr, pagour@cs.ntua.gr

Abstract. We present new, faster pseudopolynomial time algorithms for the k-Subset Sum problem, defined as follows: given a set Z of n positive integers and k targets $t_1, \ldots, t_k$, determine whether there exist k disjoint subsets $Z_1, \ldots, Z_k \subseteq Z$, such that $\Sigma(Z_i) = t_i$, for $i = 1, \ldots, k$. Assuming $t = \max\{t_1, \ldots, t_k\}$ is the maximum among the given targets, a standard dynamic programming approach based on Bellman's algorithm [3] can solve the problem in $O(nt^k)$ time. We build upon recent advances on Subset Sum due to Koiliaris and Xu [16] and Bringmann [4] in order to provide faster algorithms for k-Subset Sum. We devise two algorithms: a deterministic one of time complexity $\tilde{O}(n^{k/(k+1)}t^k)$ and a randomised one of $\tilde{O}(n + t^k)$ complexity.

We further demonstrate how these algorithms can be used in order to cope with variations of k-Subset Sum, namely Subset Sum Ratio, k-Subset Sum Ratio and Multiple Subset Sum.

Keywords: Subset Sum · FFT · Color Coding · Multiple Subset Sum · Multiple Knapsack · k-Subset Sum · Pseudopolynomial Algorithms

1 Introduction

Subset Sum is a fundamental, extensively studied problem which has recently seen tremendous progress with respect to its pseudopolynomial time solvability. Namely, fast pseudopolynomial time algorithms for Subset Sum have been presented by Koiliaris and Xu [16] and Bringmann [4], representing the first substantial improvements over the long-standing standard approach of Bellman [3], and the improvement by Pisinger [23].

Equal Subset Sum is a less studied, nevertheless noteworthy problem which has attracted the attention of several researchers, as it finds interesting applications in computational biology [8,11], computational social choice [18], and

Research supported in part by the PEVE 2020 basic research support programme of the National Technical University of Athens.

J. Chen et al. (Eds.): IJTCS-FAW 2021, LNCS 12874, pp. 37–52, 2022.
https://doi.org/10.1007/978-3-030-97099-4_3

cryptography [24], to name a few. Moreover, it is related to important theoretical concepts such as the complexity of search problems in the class TFNP [22]. However, Equal Subset Sum can be solved in $\tilde{O}(n+t)$ time[1] (where t is a bound for the sums we are interested in) by a simple extension of Bellman's [3] algorithm, as it only asks for two disjoint subsets of equal sum s, for some $s \leq t$, and as such can exploit the *disjointness property*: if there exists a pair of sets with equal sum, then one can produce a disjoint pair of sets with equal sum by removing the common elements from the initial pair.

A more general version of the problem, which is the centrepiece of this paper, is the one that asks for k disjoint subsets the sums of which are respectively equal to targets $t_i, i = 1 \ldots k$, henceforth referred to as the k-Subset Sum problem. One can see that even in the case of $k = 2$ and $t_1 = t_2$, the problem seems to escalate in complexity; this can be seen as a targeted version of Equal Subset Sum, for which the disjointness property does not hold. An interesting special case is when the sum of targets equals the sum of the input elements, that is, we ask for a partition of the input set to subsets of particular sums; an even more special case is the one in which we want to partition the input set to a number of subsets of equal sum. The latter can find applications in fair allocation situations.

1.1 Related Work

Equal Subset Sum and its optimisation version called Subset Sum Ratio [2] are well studied problems, closely related to problems appearing in many scientific areas. Some examples are the Partial Digest problem, which comes from computational biology [8,11], the allocation of individual goods [18], tournament construction [15], and a variation of Subset Sum, namely the Multiple Integrated Sets SSP, which finds applications in the field of cryptography [24]. Moreover, it is related to important concepts in theoretical computer science; for example, a restricted version of Equal Subset Sum lies in a subclass of the complexity class TFNP, namely in PPP [22], a class consisting of search problems that always have a solution due to some pigeonhole argument, and no polynomial time algorithm is known for this restricted version.

Equal Subset Sum has been proven NP-hard by Woeginger and Yu [25] and several variations have been proven NP-hard by Cieliebak *et al.* in [9,10]. A 1.324-approximation algorithm has been proposed for Subset Sum Ratio in [25] and several FPTASs appeared in [2,19,21], the fastest so far being the one in [19] of complexity $O(n^4/\varepsilon)$.

Regarding exact algorithms, recent progress has shown that it can be solved probabilistically in $O^{(1.7088^n)}$ time [20], faster than a standard "meet-in-the-middle" approach yielding an $O^{(3^{n/2})} \leq O^{(1.7321^n)}$ time algorithm. Additionally, Equal Subset Sum can be solved exactly in pseudopolynomial $\tilde{O}(n+t)$ time using an extension of Bellman's [3] algorithm. However, these techniques do not seem to apply to k-Subset Sum, mainly because we cannot assume the minimality of the involved subsets. To the best of our knowledge, no pseudopolynomial

[1] $\tilde{O}$ notation ignores polylogarithmic factors.

time algorithm substantially faster than the standard $O(nt^k)$ dynamic programming approach was known for k-SUBSET SUM prior to this work.

These problems are tightly connected to SUBSET SUM, which has seen impressive advances recently, due to Koiliaris and Xu [16] who gave a deterministic $\tilde{O}(\sqrt{n}t)$ algorithm, where n is the number of input elements and t is the target, and by Bringmann [4] who gave a $\tilde{O}(n+t)$ randomised algorithm. Jin and Wu proposed a simpler and faster randomised algorithm [14] achieving the same bounds as [4], which however seems to only solve the decision version of the problem. In a very recent work, Bringmann and Nakos [5] have presented an $O(|\mathcal{S}_t(Z)|^{4/3} poly(\log t))$ algorithm, where $\mathcal{S}_t(Z)$ is the set of all subset sums of the input set Z that are smaller than t, based on top-k convolution.

MULTIPLE SUBSET SUM is a special case of MULTIPLE KNAPSACK, both of which have attracted considerable attention. Regarding MULTIPLE SUBSET SUM, Caprara *et al.* present a PTAS for the case where all targets are the same [6], and subsequently in [7] they introduce a 3/4 approximation algorithm. The MULTIPLE KNAPSACK problem has been more intensively studied in recent years as applications for it arise naturally (in fields such as transportation, industry, and finance, to name a few). Some notable studies on variations of the problem are given by Lahyani *et al.* [17] and Dell'Amico *et al.* [13]. Special cases and variants of MULTIPLE SUBSET SUM, such as the k-SUBSET SUM problem, have been studied in [9,10] where simple pseudopolynomial algorithms were proposed.

1.2 Our Contribution

We first present two algorithms for k-SUBSET SUM: a deterministic one of complexity $\tilde{O}(n^{k/(k+1)}t^k)$ and a randomised one of complexity $\tilde{O}(n+t^k)$. We subsequently show how these ideas can be extended to solve the decision versions of SUBSET SUM RATIO, k-SUBSET SUM RATIO and MULTIPLE SUBSET SUM.

Our algorithms extend and build upon the algorithms and techniques proposed by Koiliaris and Xu [16] and Bringmann [4] for SUBSET SUM. In particular, we make use of FFT computations, modular arithmetic and color-coding, among others.

We start by presenting some necessary background in Sect. 2. Then, we present the two k-SUBSET SUM algorithms in Sect. 3. We next show how these algorithms can be used to efficiently decide multiple related subset problems. Finally, we conclude the paper by presenting some directions for future work.

2 Preliminaries

2.1 Notation

We largely follow the notation used in [4] and [16].

- Let $[x] = \{0, \ldots, x\}$ denote the set of integers in the interval $[0, x]$.
- Given a set $Z \subseteq \mathbb{N}$, we denote:
  - the sum of its elements by $\Sigma(Z) = \sum_{z \in Z} z$.

- the *characteristic polynomial of* Z by $f_Z(x) = \sum_{z \in Z} x^z$.
- the *k-modified characteristic polynomial of* Z by $f_Z^k(\vec{x}) = \sum_{z \in Z} \sum_{i=1}^k x_i^z$, where $\vec{x} = (x_1, \ldots, x_k)$.
- the *set of all subset sums of* Z *up to* t by $\mathcal{S}_t(Z) = \{\Sigma(X) \mid X \subseteq Z\} \cap [t]$.

– For two sets $X, Y \subseteq \mathbb{N}$, let
  - $X \oplus Y = \{x + y \mid x \in X \cup \{0\}, y \in Y \cup \{0\}\}$ denote the *sumset* or *pairwise sum* of sets X and Y.
  - $X \oplus_t Y = (X \oplus Y) \cap [t]$ denote the *t-capped sumset* or *t-capped pairwise sum* of sets X and Y. Note that $t > 0$.

– The pairwise sum operations can be extended to sets of multiple dimensions. Formally, let $X, Y \subseteq \mathbb{N}^k$. Then, $X \oplus Y = \{(x_1 + y_1, \ldots, x_k + y_k)\}$, where $(x_1, \ldots, x_k) \in X \cup \{0\}^k$ and $(y_1, \ldots, y_k) \in Y \cup \{0\}^k$.

2.2 Using FFT for Subset Sum

Given two sets $A, B \subseteq \mathbb{N}$ and an upper bound $t > 0$, one can compute the t-capped pairwise sum set $A \oplus_t B$ using FFT to get the monomials of the product $f_A \cdot f_B$ that have exponent $\leq t$. This operation can be completed in time $O(t \log t)$. Observe that the coefficients of each monomial x^i represent the number of pairs (a, b) that sum up to i, where $a \in A \cup \{0\}$ and $b \in B \cup \{0\}$.

Also note that an FFT operation can be extended to polynomials of multiple variables. Thus, assuming an upper bound t for the exponents involved, it is possible to compute $f_A^k \cdot f_B^k$ in $O(t^k \log t)$ time.

Lemma 1. *Given two sets of points* $S, T \subseteq [t]^k$ *one can use multidimensional FFT to compute the set of pairwise sums (that are smaller than* t*) in time* $O(t^k \log t)$.

As shown in [12, Ch. 30], given two set of points $S, T \subseteq [t]^k$ one can pipeline k one-dimensional FFT operations in order to compute a multi-dimensional FFT in time $O(t^k \log t)$.

3 k-Subset Sum

In this section we propose algorithms that build on the techniques of Koiliaris and Xu [16] and Bringmann [4] in order to solve k-Subset Sum: given a set Z of n positive integers and k targets $t_1, \ldots, t_k$, determine whether there exist k disjoint subsets $Z_1, \ldots, Z_k \subseteq Z$, such that $\Sigma(Z_i) = t_i$, for $i = 1, \ldots, k$. Note that k is fixed and not part of the input. For the rest of this section, assume that $Z = \{z_1, \ldots, z_n\}$ is the input set, $t_1, \ldots, t_k$ are the given targets and $t = \max\{t_1, \ldots, t_k\}$.

The main challenge in our approach is the fact that the existence of subsets summing up to the target numbers (or any other pair of numbers) does not imply the disjointness of said subsets. Hence, it is essential to guarantee the disjointness of the resulting subset sums through the algorithm.

Note that one can easily extend Bellman's classic dynamic programming algorithm [3] for SUBSET SUM to solve this problem in $O(nt^k)$ time.

3.1 Solving k-SUBSET SUM in Deterministic $\tilde{O}(n^{k/(k+1)}t^k)$ Time

In this section we show how to decide k-SUBSET SUM in $\tilde{O}(n^{k/(k+1)}t^k)$ time, where $t = \max\{t_1, \ldots, t_k\}$. To this end, we extend the algorithm proposed by Koiliaris and Xu [16]. We first describe briefly the original algorithm for completeness and clarity.

The algorithm recursively partitions the input set S into two subsets of equal size and returns all pairwise sums of those sets, along with cardinality information for each sum. This is achieved using FFT for pairwise sums as discussed in Sect. 2.

Using properties of congruence classes, it is possible to further capitalise on the efficiency of FFT for subset sums as follows. Given bound t, partition the elements of the initial set into congruence classes mod b, for some integer $b > 0$. Subsequently, divide the elements by b and keep their quotient, hence reducing the size of the maximum exponent value from t to t/b. One can compute the t-capped sum of the initial elements by computing the t/b-capped sum of each congruence class and combining the results of each such operation in a final t-capped FFT operation, taking into account the remainder of each congruence class w.r.t. the number of elements (cardinality) of each subset involved. In order to achieve this, it is necessary to keep cardinality information for each sum (or monomial) involved, which can be done by adding another variable with minimal expense in complexity.

Thus, the overall complexity of FFT operations for each congruence class $l \in \{0, 1, \ldots, b-1\}$ is $O((t/b)n_l \log n_l \log t)$, where n_l denotes the number of the elements belonging to said congruence class. Combining these classes takes $O(bt \log t)$ time so the final complexity is $O(t \log t(\frac{n \log n}{b} + b))$. Setting $b = \lfloor \sqrt{n \log n} \rfloor$ gives $O(\sqrt{n \log n}\, t \log t) = \tilde{O}(\sqrt{n}\, t)$. After combining the sums in the final step, we end up with a polynomial that contains each realisable sum represented by a monomial, the coefficient of which represents the number of different (not necessarily disjoint) subsets that sum up to the corresponding sum.

Our algorithm begins by using the modified characteristic polynomials we proposed in the preliminaries, thereby representing each $z \in Z$ as $\Sigma_{i=1}^k x_i^z$ in the base polynomial at the leaves of the recursion. We also use additional dimensions for the cardinalities of the subsets involved; each cardinality is represented by the exponent of some x_i, with index i greater than k. We then proceed with using multi-variate FFT in each step of the recursion in an analogous manner as in the original algorithm, thereby producing polynomials with terms $x_1^{t'_1} \ldots x_k^{t'_k} x_{k+1}^{c_1} \ldots x_{2k}^{c_k}$, each of which corresponds to a $2k$-tuple of disjoint subsets of sums $t'_1, \ldots, t'_k$ and cardinalities $c_1, \ldots, c_k$ respectively.

This results in FFT operations on tuples of $2k$ dimensions, having k dimensions of max value t/b, and another k dimensions of max value n_l which represent the cardinalities of the involved subsets, requiring $O(n_l^k (t/b)^k \log(n_l) \log(n_l t/b))$ time for congruence class $l \in \{0, 1, \dots, b-1\}$.

The above procedure is implemented in Algorithm 1, while the main algorithm is Algorithm 2, which combines the results from each congruence class, taking additional $O(bt^k \log t)$ time.

Lemma 2. *There is a term $x_1^{t'_1} \dots x_k^{t'_k}$ in the resulting polynomial if and only if there exist k disjoint subsets $Z_1, \dots, Z_k \subseteq Z$ such that $\Sigma(Z_i) = t'_i$.*

Proof. We observe that each of the terms of the form $x_1^{t'_1} \dots x_k^{t'_k}$ has been produced at some point of the recursion via an FFT operation combining two terms that belong to different subtrees, ergo containing different elements in each subset involved. As such, $t'_1, \dots, t'_k$ are sums of disjoint subsets of Z.

Theorem 1. *Given a set $Z = \{z_1, \dots, z_n\} \subseteq \mathbb{N}$ of size n and targets $t_1, \dots, t_k$, one can decide k-SUBSET SUM in time $\tilde{O}(n^{k/(k+1)} t^k)$, where $t = \max\{t_1, \dots, t_k\}$.*

Proof. The correctness of the algorithm stems from Lemma 2.
Complexity. The overall complexity of the algorithm, stemming from the computation of subset sums inside the congruence classes and the combination of those sums, is

$$O\left(t^k \log t \left(\frac{n^k \log n}{b^k} + b\right)\right) = \tilde{O}(n^{k/(k+1)} t^k),$$

where the right-hand side is obtained by setting $b = \sqrt[k+1]{n^k \log n}$.

Algorithm 1: `DisjointSC`(S, t)

Input : A set S of n positive integers and an upper bound integer t.
Output: The set $Z \subseteq (\mathcal{S}_t(S) \times [n])^k$ of all k-tuples of subset sums occurring from disjoint subsets of S up to t, along with their respective cardinality information.

1 **if** $S = \{s\}$ **then return**
$\{0\}^{2k} \cup \{(s, \underbrace{0, \dots, 0}_{k-1}, 1, \underbrace{0, \dots, 0}_{k-1})\} \cup \dots \cup \{(\underbrace{0, \dots, 0}_{k-1}, s, \underbrace{0, \dots, 0}_{k-1}, 1)\}$
2 $T \leftarrow$ an arbitrary subset of S of size $\lfloor \frac{n}{2} \rfloor$
3 **return** `DisjointSC`$(T, t) \oplus$ `DisjointSC`$(S \setminus T, t)$

Algorithm 2: DisjointSS(Z, t)

Input : A set Z of n positive integers and an upper bound integer t.
Output: The set $S \subseteq (\mathcal{S}_t(Z))^k$ of all k-tuples of subset sums up to t occurring from disjoint subsets of Z.

1 $b \leftarrow \lfloor \sqrt[k+1]{n^k \log n} \rfloor$
2 **for** $l \in [b-1]$ **do**
3 $\quad S_l \leftarrow Z \cap \{x \in \mathbb{N} \mid x \equiv l \pmod{b}\}$
4 $\quad Q_l \leftarrow \{\lfloor x/b \rfloor \mid x \in S_l\}$
5 $\quad \mathcal{R}(Q_l) \leftarrow$ DisjointSC$(Q_l, \lfloor t/b \rfloor)$
6 $\quad R_l \leftarrow \{(z_1 b + j_1 l, \ldots, z_k b + j_k l) \mid (z_1, \ldots, z_k, j_1, \ldots, j_k) \in \mathcal{R}(Q_l)\}$
7 **end**
8 **return** $R_0 \oplus_t \cdots \oplus_t R_{b-1}$

3.2 Solving k-SUBSET SUM in Randomised $\tilde{O}(n + t^k)$ Time

We will show that one can decide k-SUBSET SUM in $\tilde{O}(n + t^k)$ time, where $t = \max\{t_1, \ldots, t_k\}$, by building upon the techniques used in [4]. In particular, we will present an algorithm that successfully detects, with high probability, whether there exist k disjoint subsets each summing up to t_i respectively. In comparison to the algorithm of [4], a couple of modifications are required, which we will first briefly summarise prior to presenting the complete algorithm.

- In our version of ColorCoding, the number of repetitions of random partitioning is increased, without asymptotically affecting the complexity of the algorithm, since it remains $O(\log(1/\delta))$.
- In our version of ColorCoding, after each partition of the elements, we execute the FFT operations on the k-modified characteristic polynomials of the resulting sets. Thus, for each element s_i we introduce k points $(s_i, 0, \ldots, 0)$, $\ldots$, $(0, \ldots, s_i)$, represented by polynomial $x_1^{s_i} + \ldots + x_k^{s_i}$. Hence ColorCoding returns a set of points, each of which corresponds to k sums, realisable by disjoint subsets of the initial set.
- Algorithm ColorCodingLayer needs no modification, except that now the FFT operations concern sets of points and not sets of integers, hence the complexity analysis differs. Additionally, the algorithm returns a set of points, each of which corresponds to realisable sums by disjoint subsets of the l-layer input instance.
- The initial partition of the original set to l-layer instances remains as is, and the FFT operations once more concern sets of points instead of sets of integers.

Small Cardinality Solutions. We will first describe the modified procedure ColorCoding for solving k-SUBSET SUM if the solution size is small, i.e., an algorithm that, given an integer c, finds k-tuples of sums $(\Sigma(Y_1), \ldots, \Sigma(Y_k))$, where $\Sigma(Y_i) \leq t$ and $Y_i \subseteq Z$ are disjoint subsets of input set Z of cardinality at most c, and determines with high probability whether there exists a tuple $(t_1, \ldots, t_k)$ among them, for some given values t_i.

We randomly partition our initial set Z to c^2 subsets $Z_1, \ldots, Z_{c^2}$, i.e., we assign each $z \in Z$ to a set Z_i where i is chosen independently and uniformly at random from $\{1, \ldots, c^2\}$. We say that this random partition *splits* $Y \subseteq Z$ if $|Y \cap Z_i| \leq 1, \forall i$. If such a split occurs, the set returned by `ColorCoding` will contain[2] $\Sigma(Y)$. Indeed, by choosing the element of Y for those Z_i that $|Y \cap Z_i| = 1$ and 0 for the rest, we successfully generate k-tuples containing $\Sigma(Y)$ through the pairwise sum operations. The algorithm returns only valid sums, since no element is used more than once in each sum, because each element is assigned uniquely to a Z_i for each distinct partition.

Our intended goal is thus, for any k random disjoint subsets, to have a partition that splits them all. Such a partition allows us construct a k-tuple consisting of all their respective sums through the use of FFT. Hence, the question is to determine how many random partitions are required to achieve this with high probability.

The answer is obtained by observing that the probability to split a subset Y is the same as having $|Y|$ different balls in $|Y|$ distinct bins, when throwing $|Y|$ balls into c^2 different bins, as mentioned in [4]. Next, we compute the probability to split k random disjoint subsets $Y_1, \ldots, Y_k$ in the same partition.

The probability that a split occurs at a random partition for k random disjoint subsets $Y_1, \ldots, Y_k \subseteq Z$ is

$$\Pr[\text{all } Y_i \text{ are split}] = \prod_{i=1}^{k} \Pr[Y_i \text{ is split}] =$$

$$\frac{c^2-1}{c^2} \cdots \frac{c^2-(|Y_1|-1)}{c^2} \cdots \frac{c^2-1}{c^2} \cdots \frac{c^2-(|Y_k|-1)}{c^2} \geq$$

$$\left(\frac{c^2-(|Y_1|-1)}{c^2}\right)^{|Y_1|} \cdots \left(\frac{c^2-(|Y_k|-1)}{c^2}\right)^{|Y_k|} \geq$$

$$\left(1-\frac{1}{c}\right)^c \cdots \left(1-\frac{1}{c}\right)^c \geq \left(\frac{1}{2}\right)^2 \cdots \left(\frac{1}{2}\right)^2 = \frac{1}{4^k}$$

Hence, for $\beta = 4^k/(4^k-1)$, $r = \lceil \log_\beta(1/\delta) \rceil$ repetitions yield the desired success probability of $1-(1-1/4^k)^r \geq 1-\delta$. In other words, after r random partitions, for any k random disjoint subsets $Y_1, \ldots, Y_k$, there exists, with probability at least $1-\delta$, a partition that splits them all.

Lemma 3. *Given a set Z of positive integers, a sum bound t, a size bound $c \geq 1$ and an error probability $\delta > 0$,* `ColorCoding`*(Z, t, k, δ) returns a set S consisting of any k-tuple $(\Sigma(Y_1), \ldots, \Sigma(Y_k))$ with probability at least $1-\delta$, where $Y_1, \ldots, Y_k \subseteq Z$ are disjoint subsets with $\Sigma(Y_1), \ldots, \Sigma(Y_k) \leq t$ and $|Y_1|, \ldots, |Y_k| \leq c$, in $O(c^2 \log(1/\delta) t^k \log t)$ time.*

[2] In this context, "contain" is used to denote that $\Sigma(Y) = s_i$ for some i in a k-tuple $s = (s_1, \ldots, s_k) \in S$, where S is the resulting set of `ColorCoding`.

Algorithm 3: `ColorCoding`(Z, t, c, δ)

Input : A set Z of positive integers, an upper bound t, a size bound $c \geq 1$ and an error probability $\delta > 0$.
Output: A set $S \subseteq (\mathcal{S}_t(Z))^k$ containing any $(\Sigma(Y_1), \ldots, \Sigma(Y_k))$ with probability at least $1 - \delta$, where $Y_1, \ldots, Y_k \subseteq Z$ disjoint subsets with $\Sigma(Y_1), \ldots, \Sigma(Y_k) \leq t$ and $|Y_1|, \ldots, |Y_k| \leq c$.

1 $S \leftarrow \emptyset$
2 $\beta \leftarrow 4^k/(4^k - 1)$
3 **for** $j = 1, \ldots, \lceil \log_\beta(1/\delta) \rceil$ **do**
4   randomly partition $Z = Z_1 \cup Z_2 \cup \cdots \cup Z_{c^2}$
5   **for** $i = 1, \ldots, c^2$ **do**
6         $Z'_i \leftarrow (Z_i \times \{0\}^{k-1}) \cup \ldots \cup (\{0\}^{k-1} \times Z_i)$
7   **end**
8     $S_j \leftarrow Z'_1 \oplus_t \cdots \oplus_t Z'_{c^2}$
9     $S \leftarrow S \cup S_j$
10 **end**
11 **return** S

Proof. As we have already explained, if there exist k disjoint subsets $Y_1, \ldots, Y_k \subseteq Z$ with $\Sigma(Y_1), \ldots, \Sigma(Y_k) \leq t$ and $|Y_1|, \ldots, |Y_k| \leq c$, our algorithm guarantees that with probability at least $1 - \delta$, there exists a partition that splits them all. Subsequently, the FFT operations on the corresponding points produces the k-tuple.

Complexity. The algorithm performs $O(\log(1/\delta))$ repetitions. To compute a pairwise sum of k variables up to t, $O(t^k \log t)$ time is required. In each repetition, c^2 pairwise sums are computed. Hence, the total complexity of the algorithm is $O(c^2 \log(1/\delta) t^k \log t)$.

Solving k-SUBSET SUM for l-Layer Instances. In this part, we will prove that we can use the algorithm `ColorCodingLayer` from [4], to successfully solve k-SUBSET SUM for l-layer instances, defined below.

For $l \geq 1$, we call (Z, t) an *l-layer instance* if $Z \subseteq [t/l, 2t/l]$ or $Z \subseteq [0, 2t/l]$ and $l \geq n$. In both cases, $Z \subseteq [0, 2t/l]$ and for any $Y \subseteq Z$ with $\Sigma(Y) \leq t$, we have $|Y| \leq l$. The algorithm of [4] successfully solves the SUBSET SUM problem for l-layer instances. We will show that by calling the modified `ColorCoding` algorithm and modifying the FFT operations so that they concern sets of points, the algorithm successfully solves k-SUBSET SUM in such instances.

We will now prove the correctness of the algorithm. Let $X^1, \ldots, X^k \subseteq Z$ be disjoint subsets with $\Sigma(X^1), \ldots, \Sigma(X^k) \leq t$. By [4, Claim 3.3], we have that $\Pr[|Y_i| \geq 6\log(l/\delta)] \leq \delta/l$, where $Y_i = Y \cap Z_i$, for any $Y \subseteq Z$ with at most l elements. Hence, the probability that $|X_i^1| \leq 6\log(l/\delta)$ and $|X_i^2| \leq 6\log(l/\delta)$

Algorithm 4: `ColorCodingLayer`(Z, t, l, δ)

Input : An l-layer instance (Z, t) and an error probability $\delta \in (0, 1/2^{k+1}]$.
Output: A set $S \subseteq (\mathcal{S}_t(Z))^k$ containing any $(\Sigma(Y_1), \ldots, \Sigma(Y_k))$ with probability at least $1 - \delta$, where $Y_1, \ldots, Y_k \subseteq Z$ disjoint subsets with $\Sigma(Y_1), \ldots, \Sigma(Y_k) \leq t$.

1 **if** $l < \log(l/\delta)$ **then return** `ColorCoding`(Z, t, l, δ)
2 $m \leftarrow l/\log(l/\delta)$ rounded up to the next power of 2
3 randomly partition $Z = Z_1 \cup Z_2 \cup \cdots \cup Z_m$
4 $\gamma \leftarrow 6\log(l/\delta)$
5 **for** $j = 1, \ldots, m$ **do**
6 | $S_j \leftarrow$ `ColorCoding`$(Z_j, 2\gamma t/l, \gamma, \delta/l)$
7 **end**
8 **for** $h = 1, \ldots, \log m$ **do** `// combine` S_j `in a binary-tree-like way`
9 | **for** $j = 1, \ldots, m/2^h$ **do**
10 | | $S_j \leftarrow S_{2j-1} \oplus_{2^h \cdot 2\gamma t/l} S_{2j}$
11 | **end**
12 **end**
13 **return** $S_1 \cap [t]^k$

and so on, for all $i = 1, \ldots, m$ is

$$\Pr[\bigwedge_{i=1}^{m}(|X_i^1| \leq 6\log(l/\delta) \wedge \ldots \wedge |X_i^k| \leq 6\log(l/\delta))] \geq$$

$$1 - \sum_{i=1}^{m}\left(\Pr[|X_i^1| > 6\log(l/\delta)]\right) - \ldots - \sum_{i=1}^{m}\left(\Pr[|X_i^k| > 6\log(l/\delta)]\right) \geq 1 - km\delta/l.$$

`ColorCoding` computes $(\Sigma(X_i^1), \ldots, \Sigma(X_i^k))$ with probability at least $1-\delta$. This happens for all i with probability at least $1-m\delta/l$. Then, combining the resulting sets indeed yields the k-tuple $(\Sigma(X^1), \ldots, \Sigma(X^k))$. The total error probability is at most $(k+1)m\delta/l$. Assume that $\delta \in (0, 1/2^{k+1}]$. Since $l \geq 1$ and $\delta \leq 1/2^{k+1}$, we have $\log(l/\delta) \geq (k+1)$. Hence, the total error probability is bounded by δ. This gives the following.

Lemma 4. *Given an l-layer instance (Z, t), upper bound t and error probability $\delta \in (0, 1/2^{k+1}]$,* `ColorCodingLayer`$(Z, t, l, \delta)$ *solves k-*SUBSET SUM *with sum at most t in time $O\left(t^k \log t \frac{\log^{k+2}(l/\delta)}{l^{k-1}}\right)$ with probability at least $1 - \delta$.*

Complexity. The time to compute the sets of k-tuples $S_1, \ldots, S_m$ by calling `ColorCoding` is

$$O(m\gamma^2 \log(l/\delta)(\gamma t/l)^k \log(\gamma t/l)) = O\left(\frac{\gamma^{k+2}}{l^{k-1}} t^k \log t\right) = O\left(\frac{\log^{k+2}(l/\delta)}{l^{k-1}} t^k \log t\right).$$

Combining the resulting sets costs

$$O\left(\sum_{h=1}^{\log m} \frac{m}{2^h}(2^h\gamma t/l)^k \log(2^h\gamma t/l)\right) = O\left(\sum_{h=1}^{\log m} \frac{2^{h(k-1)}}{m^{k-1}} t^k \log t\right) =$$

$$O\left(\frac{t^k \log t}{m^{k-1}} \sum_{h=1}^{\log m} \left(2^{k-1}\right)^h\right) = O\left(\frac{t^k \log t}{m^{k-1}} \left(2^{k-1}\right)^{\log m}\right) = O(t^k \log t),$$

since for $c > 1$, we have that $O\left(\sum_{k=0}^{n} c^k\right) = O(c^n)$, which is dominated by the computation of $S_1, \ldots, S_m$.

Hence, `ColorCodingLayer` has total complexity $O\left(t^k \log t \cdot \frac{\log^{k+2}(l/\delta)}{l^{k-1}}\right)$.

General Case. It remains to show that for every instance (Z, t), we can construct l-layer instances and take advantage of `ColorCodingLayer` in order to solve k-SUBSET SUM for the general case. This is possible by partitioning set Z at $t/2^i$ for $i = 1, \ldots, \lceil \log n \rceil - 1$. Thus, we have $O(\log n)$ l-layers $Z_1, \ldots, Z_{\lceil \log n \rceil}$. On each layer we run `ColorCodingLayer`, and then we combine the results using pairwise sums.

Algorithm 5: `kSubsetSum`(Z, δ, t)

Input : A set of positive integers Z, an upper bound t and an error probability δ.

Output: A set $S \subseteq (\mathcal{S}_t(Z))^k$ containing any $(\Sigma(Y_1), \ldots, \Sigma(Y_k))$ with probability at least $1 - \delta$, where $Y_1, \ldots, Y_k \subseteq Z$ disjoint subsets with $\Sigma(Y_1), \ldots, \Sigma(Y_k) \leq t$.

```
1 partition Z into Z_i ← Z ∩ (t/2^i, t/2^(i-1)] for i = 1,...,⌈log n⌉ − 1 and
    Z_⌈log n⌉ ← Z ∩ [0, t/2^(⌈log n⌉−1)]
2 S ← ∅
3 for i = 1,...,⌈log n⌉ do
4   |  S_i ← ColorCodingLayer(Z_i, t, 2^i, δ/⌈log n⌉)
5   |  S ← S ⊕_t S_i
6 end
7 return S
```

We will now prove the main theorem of this section.

Theorem 2. *Given a set $Z \subseteq \mathbb{N}$ of size n and targets $t_1, \ldots, t_k$, one can decide k-SUBSET SUM in $\tilde{O}(n + t^k)$ time w.h.p., where $t = \max\{t_1, \ldots, t_k\}$.*

Proof. Let $X^1, \ldots, X^k \subseteq Z$ be k disjoint subsets with $\Sigma(X^1), \ldots, \Sigma(X^k) \leq t$, and $X_i^j = X^j \cap Z_i$, for $j = 1, \ldots, k$ and $i = 1, \ldots, \lceil \log n \rceil$. Each call to `ColorCodingLayer` returns a k-tuple $(\Sigma(X_i^1), \ldots, \Sigma(X_i^k))$ with probability at least $1-\delta/\lceil\log n\rceil$, hence the probability that all calls return the corresponding k-tuple is

$$\Pr[\texttt{ColorCodingLayer} \text{ returns } (\Sigma(X_i^1),\ldots,\Sigma(X_i^k)),\forall i] =$$

$$1-\Pr[\text{some call fails}] \geq 1-\sum_{i=1}^{\lceil\log n\rceil}\frac{\delta}{\lceil\log n\rceil} = 1-\lceil\log n\rceil\cdot\frac{\delta}{\lceil\log n\rceil} = 1-\delta$$

If all calls return the corresponding k-tuple, the algorithm successfully constructs the k-tuple $(\Sigma(X^1),\ldots,\Sigma(X^k))$. Thus, with probability at least $1-\delta$, the algorithm solves k-Subset Sum.

Complexity. Reading the input requires $\Theta(n)$ time. In each of the $\Theta(\log n)$ repetitions of the algorithm, we make a call to `ColorCodingLayer`, plus compute a pairwise sum. The computation of the pairwise sum requires $O(t^k\log t)$ time since it concerns k-tuples. For each call to `ColorCodingLayer`, we require $O\left(t^k\log t\frac{\log^{k+2}\left(\frac{2^i\log n}{\delta}\right)}{2^{i(k-1)}}\right)$ time. Hence, the overall complexity is

$$O\left(n+t^k\log t\log n+\sum_{i=1}^{\log n}t^k\log t\frac{\log^{k+2}\left(\frac{2^i\log n}{\delta}\right)}{2^{i(k-1)}}\right)=\tilde{O}(n+t^k).$$

4 Faster Algorithms for Multiple Subset Problems

The techniques developed in the previous section can be further applied to give faster pseudopolynomial algorithms for the decision version of the problems Subset Sum Ratio, k-Subset Sum Ratio and Multiple Subset Sum. In this section we will present how these algorithms can be used to efficiently solve these problems.

The algorithms we previously presented result in a polynomial $P(x_1,\ldots,x_k)$ consisting of terms each of which corresponds to a k-tuple of realisable sums by disjoint subsets of the initial input set Z. In other words, if there exists a term $x_1^{s_1}\ldots x_k^{s_k}$ in the resulting polynomial, then there exist disjoint subsets $Z_1,\ldots,Z_k\subseteq Z$ such that $\Sigma(Z_1)=s_1,\ldots,\Sigma(Z_k)=s_k$.

It is important to note that, while the deterministic algorithm of Subsect. 3.1 returns a polynomial consisting of *all* terms corresponding to such k-tuples of realisable sums by disjoint subsets, the randomised algorithm of Subsect. 3.2 does not. However, that does not affect the correctness of the following algorithms, since it suffices to guarantee that the k-tuple corresponding to the optimal solution of the respective (optimisation) problem is included with high probability. That indeed happens, since the resulting polynomial consists of *any* viable term with high probability, as discussed previously.

Subset Sum Ratio. The first variation we will discuss is Subset Sum Ratio, which asks to determine, given a set $Z\subseteq\mathbb{N}$ of size n and an upper bound t, what is the smallest ratio of sums between any two disjoint subsets $S_1, S_2 \subseteq Z$, where $\Sigma(S_1), \Sigma(S_2) \leq t$. This can be solved in deterministic $\tilde{O}(n^{2/3}t^2)$ time using the algorithm proposed in Subsect. 3.1 by simply iterating over the terms of the final polynomial that involve both parameters x_1 and x_2 and checking the ratio of their exponents. SUBSET SUM RATIO can also be solved with high probability in randomised $\tilde{O}(n+t^2)$ time using the algorithm proposed in Subsect. 3.2 instead.

k-SUBSET SUM RATIO. An additional extension is the k-SUBSET SUM RATIO problem, which asks, given a set $Z \subseteq \mathbb{N}$ of size n and k bounds $t_1, \ldots, t_k$, to determine what is the smallest ratio between the largest and smallest sum of any set of k disjoint subsets $Z_1, \ldots, Z_k \subseteq Z$ such that $\Sigma(Z_i) \leq t_i$. Similar to k-SUBSET SUM, an interesting special case is when all t_i's are equal, in which case we search for k subsets that are as similar as possible in terms of sum.

Similarly, we can solve this in deterministic $\tilde{O}(n^{k/(k+1)} \cdot t^k)$ or randomised $\tilde{O}(n+t^k)$ time by using the corresponding algorithm to solve k-SUBSET SUM and subsequently iterating over the terms of the resulting polynomial that respect the corresponding bounds, and finally evaluating the ratio of the largest to smallest exponent.

k-PARTITION. A special case of k-SUBSET SUM RATIO is the k-PARTITION problem, which asks, given a set $Z \subseteq \mathbb{N}$ of size n, to partition its elements into k subsets $Z_1, \ldots, Z_k$, while minimising the ratio among the sums $\Sigma(Z_i)$. In order to solve this problem, it suffices to solve k-SUBSET SUM RATIO for $t = \max(Z) + (\Sigma(Z)/k)$, where $\max(Z)$ denotes the largest element of Z, while only considering the terms $x_1^{s_1} \cdots x_k^{s_k}$ of the final polynomial for which $\sum s_i = \Sigma(Z)$.

MULTIPLE SUBSET SUM. Finally, we consider the MULTIPLE SUBSET SUM problem, that asks, given a set $Z \subseteq \mathbb{N}$ of size n and k bounds $t_1, \ldots, t_k$, to determine what is the maximum sum of sums of any set of k disjoint subsets $Z_1, \ldots, Z_k$ of Z, such that $\Sigma(Z_i) \leq t_i$. This problem is a special case of the MULTIPLE KNAPSACK problem and can also be seen as a generalisation of k-SUBSET SUM. It should be clear that the same techniques as those e.g. used for k-SUBSET SUM RATIO apply directly, leading to the same time complexity bounds of $\tilde{O}(n^{k/(k+1)} \cdot t^k)$ deterministically and $\tilde{O}(n + t^k)$ probabilistically.

5 Future Work

Tighter Complexity Analysis. The algorithm of Sect. 3.1 involves the computation of the possible sums by disjoint subsets along with their respective cardinality. To do so, we extend the FFT operations to multiple variables, each representing either a possible subset sum or its cardinality. Hence, for k subsets and n elements, we proceed with FFT operations on variables $x_1, \ldots, x_k, c_1, \ldots, c_k$, where the exponents of x_i are in $[t]$ for some given upper bound t, whereas the exponents of c_i in $[n]$. Notice however that each element is used only on a single subset, hence for a term $x_1^{s_1} \cdots x_k^{s_k} c_1^{n_1} \cdots c_k^{n_k}$ of the produced polynomial, it holds that $s_i \leq t$ and $\sum n_i \leq n$. This differs substantially from our analysis, where we essentially only assume that $n_i \leq n$, which is significantly less strict. Hence, a stricter complexity analysis may be possible on those FFT operations, resulting in a more efficient overall complexity for this algorithm.

Also notice that, in our proposed algorithms, each combination of valid sums appears $k!$ times. This means that for every k disjoint subsets $S_1, \ldots, S_k$ of the input set, there are $k!$ different terms in the resulting polynomial of the algorithm representing the combination of sums $\Sigma(S_1), \ldots, \Sigma(S_k)$. This massive increase on the number of terms does not influence the asymptotic analysis of our algorithms, nevertheless can be restricted for better performance. Additionally, one can limit the FFT operations to different bounds for each variable, resulting in slightly improved complexity analysis without changing the algorithms whatsoever. In this paper, we preferred to analyse the complexity of the algorithms using $t = \max\{t_1, \ldots, t_k\}$ for the sake of simplicity, but one can alternatively obtain time complexities of $\tilde{O}(n^{k/(k+1)}T)$ and $\tilde{O}(n + T)$ for the deterministic and the randomised algorithm respectively, where $T = \prod t_i$.

Recovery of Solution Sets. The algorithms introduced in this paper solve the *decision version* of the k-Subset Sum problem. In other words, their output is a binary response, indicating whether there exist k disjoint subsets whose sums are equal to given values $t_1, \ldots, t_k$ respectively. An extension of these algorithms could involve the reconstruction of the k solution subsets. Koiliaris and Xu [16] argue that one can reconstruct the solution set of Subset Sum with only polylogarithmic overhead. That is possible by carefully extracting the *witnesses* of each sum every time an FFT operation is happening. These witnesses are actually the partial sums used to compute the new sum. Thus, by reducing this problem to the *reconstruction problem* as mentioned in [1], they conclude that it is possible to compute all the witnesses of an FFT operation without considerably increasing the complexity. That is the case for one-dimensional FFT operations involving a single variable, so it may be possible to use analogous arguments for multiple variables.

Extension of Subset Sum *Algorithm Introduced by Jin and Wu.* Jin and Wu introduced an efficient $\tilde{O}(n + t)$ randomised algorithm for solving Subset Sum in [14]. This algorithm is much simpler than Bringmann's and actually has slightly better complexity. It is interesting to research whether this algorithm can be extended to cope with k-Subset Sum (and the variations mentioned in Sect. 4), as was the case for Bringmann's, since that would result in a simpler alternative approach.

References

1. Aumann, Y., Lewenstein, M., Lewenstein, N., Tsur, D.: Finding witnesses by peeling. ACM Trans. Algorithms **7**(2) (2011). https://doi.org/10.1145/1921659.1921670
2. Bazgan, C., Santha, M., Tuza, Z.: Efficient approximation algorithms for the SUBSET-SUMS EQUALITY problem. J. Comput. Syst. Sci. **64**(2), 160–170 (2002). https://doi.org/10.1006/jcss.2001.1784

3. Bellman, R.E.: Dynamic Programming. Princeton University Press, Princeton (1957)
4. Bringmann, K.: A near-linear pseudopolynomial time algorithm for subset sum. In: Proceedings of the Twenty-Eighth Annual ACM-SIAM Symposium on Discrete Algorithms, SODA 2017, pp. 1073–1084. SIAM (2017). https://doi.org/10.1137/1.9781611974782.69
5. Bringmann, K., Nakos, V.: Top-k-convolution and the quest for near-linear output-sensitive subset sum. In: Proceedings of the 52nd Annual ACM SIGACT Symposium on Theory of Computing, STOC 2020, pp. 982–995. ACM (2020). https://doi.org/10.1145/3357713.3384308
6. Caprara, A., Kellerer, H., Pferschy, U.: A PTAS for the multiple subset sum problem with different knapsack capacities. Inf. Process. Lett. **73**(3–4), 111–118 (2000). https://doi.org/10.1016/S0020-0190(00)00010-7
7. Caprara, A., Kellerer, H., Pferschy, U.: A 3/4-approximation algorithm for multiple subset sum. J. Heuristics **9**(2), 99–111 (2003). https://doi.org/10.1023/A:1022584312032
8. Cieliebak, M., Eidenbenz, S.: Measurement errors make the partial digest problem NP-hard. In: Farach-Colton, M. (ed.) LATIN 2004. LNCS, vol. 2976, pp. 379–390. Springer, Heidelberg (2004). https://doi.org/10.1007/978-3-540-24698-5_42
9. Cieliebak, M., Eidenbenz, S., Pagourtzis, A.: Composing equipotent teams. In: Lingas, A., Nilsson, B.J. (eds.) FCT 2003. LNCS, vol. 2751, pp. 98–108. Springer, Heidelberg (2003). https://doi.org/10.1007/978-3-540-45077-1_10
10. Cieliebak, M., Eidenbenz, S.J., Pagourtzis, A., Schlude, K.: On the complexity of variations of equal sum subsets. Nord. J. Comput. **14**(3), 151–172 (2008)
11. Cieliebak, M., Eidenbenz, S., Penna, P.: Noisy data make the partial digest problem NP-hard. In: Benson, G., Page, R.D.M. (eds.) WABI 2003. LNCS, vol. 2812, pp. 111–123. Springer, Heidelberg (2003). https://doi.org/10.1007/978-3-540-39763-2_9
12. Cormen, T.H., Leiserson, C.E., Rivest, R.L., Stein, C.: Introduction to Algorithms, 3rd edn. MIT Press, Cambridge (2009)
13. Dell'Amico, M., Delorme, M., Iori, M., Martello, S.: Mathematical models and decomposition methods for the multiple knapsack problem. Eur. J. Oper. Res. **274**(3), 886–899 (2019). https://doi.org/10.1016/j.ejor.2018.10.043
14. Jin, C., Wu, H.: A simple near-linear pseudopolynomial time randomized algorithm for subset sum. In: 2nd Symposium on Simplicity in Algorithms (SOSA 2019), vol. 69, pp. 17:1–17:6 (2018). https://doi.org/10.4230/OASIcs.SOSA.2019.17
15. Khan, M.A.: Some problems on graphs and arrangements of convex bodies (2017). https://doi.org/10.11575/PRISM/10182
16. Koiliaris, K., Xu, C.: Faster pseudopolynomial time algorithms for subset sum. ACM Trans. Algorithms **15**(3) (2019). https://doi.org/10.1145/3329863
17. Lahyani, R., Chebil, K., Khemakhem, M., Coelho, L.C.: Matheuristics for solving the multiple knapsack problem with setup. Comput. Ind. Eng. **129**, 76–89 (2019). https://doi.org/10.1016/j.cie.2019.01.010
18. Lipton, R.J., Markakis, E., Mossel, E., Saberi, A.: On approximately fair allocations of indivisible goods. In: Proceedings 5th ACM Conference on Electronic Commerce (EC-2004), pp. 125–131. ACM (2004). https://doi.org/10.1145/988772.988792
19. Melissinos, N., Pagourtzis, A.: A faster FPTAS for the subset-sums ratio problem. In: Wang, L., Zhu, D. (eds.) COCOON 2018. LNCS, vol. 10976, pp. 602–614. Springer, Cham (2018). https://doi.org/10.1007/978-3-319-94776-1_50

20. Mucha, M., Nederlof, J., Pawlewicz, J., Wegrzycki, K.: Equal-subset-sum faster than the meet-in-the-middle. In: 27th Annual European Symposium on Algorithms, ESA 2019. LIPIcs, vol. 144, pp. 73:1–73:16 (2019). https://doi.org/10.4230/LIPIcs.ESA.2019.73
21. Nanongkai, D.: Simple FPTAS for the subset-sums ratio problem. Inf. Process. Lett. **113**(19–21), 750–753 (2013). https://doi.org/10.1016/j.ipl.2013.07.009
22. Papadimitriou, C.H.: On the complexity of the parity argument and other inefficient proofs of existence. J. Comput. Syst. Sci. **48**(3), 498–532 (1994). https://doi.org/10.1016/S0022-0000(05)80063-7
23. Pisinger, D.: Linear time algorithms for knapsack problems with bounded weights. J. Algorithms **33**(1), 1–14 (1999). https://doi.org/10.1006/jagm.1999.1034
24. Voloch, N.: MSSP for 2-D sets with unknown parameters and a cryptographic application. Contemp. Eng. Sci. **10**, 921–931 (2017). https://doi.org/10.12988/ces.2017.79101
25. Woeginger, G.J., Yu, Z.: On the equal-subset-sum problem. Inf. Process. Lett. **42**(6), 299–302 (1992). https://doi.org/10.1016/0020-0190(92)90226-L

Hardness and Algorithms for Electoral Manipulation Under Media Influence

Liangde Tao[1], Lin Chen[2(✉)], Lei Xu[3], Shouhuai Xu[4], Zhimin Gao[5], Weidong Shi[5], and Dian Huang[6,7]

[1] Department of Computer Science, Zhejiang University, Hangzhou, China
[2] Department of Computer Science, Texas Tech University, Lubbock, USA
[3] Department of Computer Science, University of Texas Rio Grande Valley, Edinburg, USA
[4] Department of Computer Science, University of Colorado Colorado Springs, Colorado Springs, USA
sxu@uccs.edu
[5] Department of Computer Science, University of Houston, Houston, USA
gao@kell.vin
[6] College of Management and Economics, Tianjin University, Tianjin, China
huangdian@tju.edu.cn
[7] School of New Media and Communication, Tianjin University, Tianjin, China

Abstract. In this paper, we study a generalization of the classic bribery problem known as electoral manipulation under media influence (EMMI). This model is motivated by modern political campaigns where candidates try to convince voters through advertising in media (TV, newspaper, Internet). When compared with the classical bribery problem, the attacker in this setting cannot directly change opinions of individual voters, but instead can execute influences via a set of manipulation strategies (e.g., advertising on a TV channel). Different manipulation strategies incur different costs and influence different subsets of voters. Once receiving a significant amount of influence, a voter will change opinion. To characterize the opinion change of each voter, we adopt the well-accepted threshold model. We prove the NP-hardness of the EMMI problem and give a dynamic programming algorithm that runs in polynomial time for a restricted case of the EMMI problem.

Keywords: Computational Social Choice · Computational Complexity · Dynamic Programming

1 Introduction

Election manipulation is an important topic in the field of *computational social choice*, which has a rich body of literature including approximation algorithms and computational complexity of variants of the problem (see, e.g., [1]). However, most studies focus on "direct" electoral manipulations, namely that a briber may alter the opinion of a voter at some cost, where the term *opinion* is used interchangeably with the term *preference* throughout the paper. For example, in the seminal work [12], a briber can pay a fixed (but voter-dependent) amount of money to alter a voter's preference arbitrarily.

J. Chen et al. (Eds.): IJTCS-FAW 2021, LNCS 12874, pp. 53–64, 2022.
https://doi.org/10.1007/978-3-030-97099-4_4

In this paper, we study elections under "indirect" manipulations, leading to a new problem known as electoral manipulation under media influence (EMMI). In this setting, various kinds of media are important means for people to retrieve information. Therefore, advertising in media is widely adopted for political campaigns. A significant amount of money has been spent to spread information through TV, Internet, newspaper, etc., to sway or reinforce the opinion of voters during a large-scale political campaign. For example, Facebook reports that $400 million was spent on its platform for political advertisements between May of 2018 and the Election Day, ranging from U.S Senate to county sheriff races [15]. In sharp contrast to the classical bribery problem where a candidate may directly exchange money for a vote from a voter, this "indirect" way of manipulation may target all of the voters and succeed with some of them.

To characterize how a voter's opinion or preference is influenced by the information received from media advertisements, by adopting the *threshold model* that has been widely accepted in the research field of social networks (in particular influence maximization). In the threshold model, each voter has a fixed threshold. Each piece of information a voter received from a certain media source incurs a certain amount of influence to the voter. When the total amount of influence a voter receives reaches the threshold, the voter will change opinion. Because election candidates can use advertisement or manipulation strategies to target specific groups of voters, we use a manipulation strategy to represent an "indirect" way of casting influence over a subset of voters (e.g., advertising on a local newspaper vs. national TV channel). Different manipulation strategies may incur different costs. As mentioned above, whether or not a voter will change opinion depends on whether the received total amount of influence from the set of manipulation strategies reaches the voter's threshold or not. Similar to the classic bribery problem, we assume there is an attacker who attempts to make a designated candidate win. The attacker has a fixed budget for manipulations and can choose from a set of potential manipulation strategies. Each manipulation strategy incurs some cost and can influence a subset of voters. The goal of the attacker is to select a subset of strategies within the budget so that a designated candidate can win.

Our model is motivated by considering the flow of information from media to voters. Studying the electoral manipulation problem under this model would be helpful to explain what is going on in, e.g., political campaigns where candidates attempt to convince voters through various kinds of media venues. Throughout this paper, we will focus on one of the most fundamental voting rules known as *plurality*, namely that each voter votes for its most preferred candidate, and the winner is the candidate receiving the highest number of votes.

Our Contributions. The main contribution of this paper is in two-fold, *conceptual* and *technical*. The conceptual contribution is the introduction of the EMMI problem. This problem generalizes the setting of the classical bribery problem by taking into account sophisticated and realistic means for electoral manipulations, namely changing voters' opinions via media influence. We further model manipulation strategies, which incur different costs and execute influences on different subsets of voters. A voter usually receives information from multiple sources and may only switch in favor of a designated candidate if a significant number of sources convince the voter to do so. Inspired by the research in social networks, we adopt the threshold model to characterize how a voter may be influenced by the manipulation strategies.

The technical contribution is the proof of the NP-hardness of the EMMI problem. We find that the structure of the influence sets of the strategies, which characterizes the connections between strategies and voters, dramatically affects the hardness of the problem. In the general case (i.e., there are no constraints on this structure), we prove that the EMMI problem is NP-hard even if there are only two candidates and each strategy incurs a unit cost. In real life, there are usually constraints on this structure. For example, the laminar structure reflects the hierarchy of strategies; e.g., in the case of TV channels, the influence set of a local TV channel is likely contained in a larger scale nationwide TV channel. Under this structural constraint with a constant number of candidates, we give a dynamic programming algorithm for the EMMI problem which runs in polynomial time.

Related Work. Our problem is closely related with the bribery problem in computational social choice. Faliszewski et al. [12] introduced the fundamental bribery problem. Following their work, various extensions of the bribery problem has been considered in the literature; see, e.g., [2–4,7,9,10,13,14,20,23]. In particular, there is a trend to study the electoral control problem on the social network [22]. In their model, the attacker bribes an initial subset of voters. Every bribed voter will influence his/her neighbor voters. Once an unbribed voter receives the influences more than the threshold, the voter will be bribed and continue to influence his/her neighbor voters. Such propagation will continue until no more voters are bribed. They assume the bribing cost for each voter is unit. Extensive researches have been carried out for this problem see, e.g., [5,8,11,19]. We remark that in our model the influence only goes from information providers, as specified by different manipulation strategies, to the voters. More importantly, the cost of adopting different strategies varies in our model which is more general.

Our problem is also related to the set multicover problem (i.e., take a vote as an element and a strategy as a set). The set multicover problem is a well-known optimization problem and various extensions have been considered in the literature; see, e.g., [6,16,17,21]. We want to emphasize that there is a huge difference between the set multicover problem and the EMMI problem. In the set multicover problem, the covering constraint for each element (voter) is given by the input. Whereas, in the EMMI problem, we need to decide two things: the total number of voters we need to bribe from each candidate; the specific voter we need to bribe.

Paper Outline. Section 2 formalizes the EMMI problem. Section 3 proves its hardness in the general setting. Section 4 presents a dynamic programming for the EMMI problem in the special setting mentioned above. Section 5 concludes the paper.

2 Problem Formalization

More precisely, there are a set of m candidates $C = \{c_1, c_2, \ldots, c_m\}$ and a set of n voters $V = \{v_1, v_2, \ldots, v_n\}$. The preference of each voter is known in the absence of the bribery. There is additionally a set of r manipulation strategies $S = \{s_1, s_2, \cdots, s_r\}$ for an attacker. Each manipulation strategy s_h is associated with a cost $q_h \in \mathbb{Z}_+$ and an

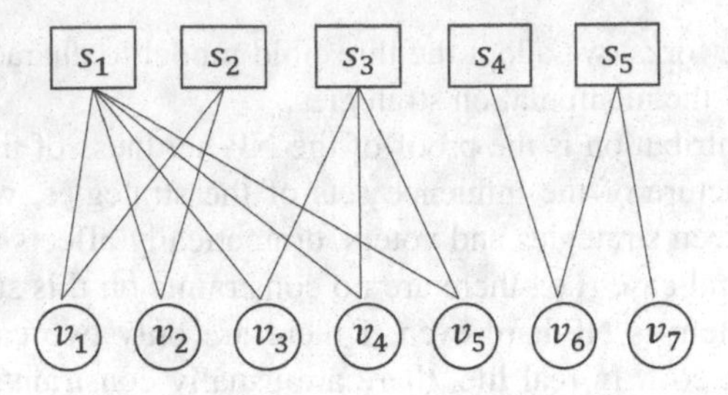

Fig. 1. An example of the EMMI problem

influence set $I_h \subseteq V$, which is the subset of voters this strategy can influence. For ease of presentation, we can establish a bipartite graph $G = (S \cup V, E)$, where by abusing the notation a bit we denote its vertices by s_h and v_j. Each s_h is connected to v_j's that belong to its influence set I_h. Figure 1 illustrates an instance of the EMMI problem with 5 strategies $(s_1, \ldots, s_5)$ and 7 voters $(v_1, \ldots, v_7)$. For example, the second strategy s_2 influences 2 voters, namely v_1, v_2.

In our model, we assume the threshold of voter v_j is T_j which is known. If the total amount of influence received by voter v_j (i.e., the number of selected strategies whose influence set contains v_j) is larger than or equal to T_j, then v_j will change preference and vote for the designated candidate; otherwise, v_j will not change the preference. Without loss of generality, we assume c_1 is the winner in the absence of manipulation, and c_m is the candidate preferred by the attacker. Let $V_i \subseteq V$ be the subset of voters who votes for v_i in the absence of manipulation. We may re-index the candidates such that $|V_1| \geq |V_2| \geq \cdots \geq |V_{m-1}|$. The goal of the attacker is to select a subset of strategies within a given budget such that c_m becomes the winner under the plurality rule.

Using the preceding notations, the EMMI problem is formalized as follows:

The EMMI Problem
Input: A set of m candidates $C = \{c_1, c_2, \ldots, c_m\}$, where c_1 is the winner in the absence of manipulation and c_m is the designated candidate; a set of n voters $V = \{v_1, v_2, \cdots, v_n\}$ with $V = \cup_{i=1}^{m} V_i$, where V_i is the subset of voters that vote for c_i in the absence of manipulation, and $|V_1| \geq |V_2| \geq \cdots \geq |V_{m-1}|$; a threshold T_j for every v_j; the attacker's budget B; a set of r manipulation strategies $S = \{s_1, s_2, \cdots, s_r\}$, where each $s_h \in S$ is associated with a cost q_h and an influence set I_h; every pair (s_h, v_j) where $v_j \in I_h$ can bring a unit amount of influence to v_j; a voter (who does not vote for c_m originally) will vote for c_m if $\sigma_j \geq T_j$, where σ_j is the total number of selected strategies whose influence set contains v_j.
Output: Does there exists a subset of strategies $S^* \subseteq S$ which can let the designated candidate be the winner under the plurality rule while satisfying the budget constraint (e.g., $\sum_{s_h \in S^*} q_h \leq B$)?

We remark that our EMMI problem is general enough to incorporate the classical bribery problem, which is studied extensively in the literature, as a special case. In the bribery problem, the attacker can pay some cost to change one voter's opinion. This can be viewed as a special case of the EMMI problem where each strategy only covers one distinct voter and the threshold of every voter is 1.

3 Strong Intractability for the EMMI Problem

The goal of this section is to prove the NP-hardness of the EMMI problem. In fact, we have a much stronger hardness result for the EMMI problem. Specifically, we have the following theorem.

Theorem 1. *Even if there are only two candidates and all the strategies have the same cost, there is no FPT algorithm parameterized by $k = |V_1| - |V_m|$ and $T = \max_j T_j$ for the EMMI problem, unless FPT= W[1].*

In order to prove Theorem 1, we leverage the strong intractability result for the bi-clique problem in [18], which is reviewed below.

Theorem 2 *[18]. There is a polynomial time algorithm $\mathbb{A}$ such that for every graph G with n vertices and $\gamma \in \mathbb{Z}_+$ with $\lceil n^{\frac{\gamma}{\gamma+6}} \rceil > (\gamma+1)!$ and $6|(\gamma+1)$, the algorithm $\mathbb{A}$ constructs a bipartite graph $H = (A \dot{\cup} B, E)$ satisfying:*

(i) If G contains a clique of size γ, then there are α vertices in A with at least $\lceil n^{\frac{\gamma}{\gamma+6}} \rceil$ common neighbors in B;
(ii) If G does not contain a clique of size γ, every α vertices in A have at most $(\gamma+1)!$ common neighbors in B,

where $\alpha = \binom{\gamma}{2}$.

Briefly speaking, Theorem 2 shows a gap reduction from k-CLIQUE to k-BICLIQUE, and hence rules out the existence of an FPT algorithm for k-BICLIQUE, assuming FPT $\neq$ W[1]. To prove our Theorem 1, we proceed by contradiction. We assume, on the contrary, that an FPT time algorithm for EMMI under the deterministic threshold model exists, and use that to test whether A contains α vertices who share a large number of neighbors in the constructed bipartite graph H.

Proof (of Theorem 1). Suppose on the contrary that there exists an FPT algorithm parameterized by $k = |V_1| - |V_m|$ and $T = \max_j T_j$, we prove that there exists an FPT algorithm, parameterized by γ, for determining whether a given graph G admits a clique of size γ, which contradicts FPT $\neq$ W[1].

Given a graph G, which is an instance of γ-CLIQUE, we construct an instance of the EMMI problem as follows. We first construct the bipartite graph $H = (A \cup B, E)$ using Theorem 2. There are only two candidates c_1 and c_2, with c_1 being the original winner and c_2 being the voter preferred by the attacker. We let A be the set of all strategies, each of unit cost, and B be the set of key voters. Every key voter votes for c_1. In addition to the key voters, we construct a suitable number of dummy voters. Dummy voters will not be influenced by any strategy, and will only vote for c_2. We set the number of dummy voters $(\gamma+1)!$ such that by counting all the votes from the dummy and key voters, c_2 receives exactly $2(\gamma+1)!+1$ votes less than c_1, i.e., $k = 2(\gamma+1)!+1$. That means, c_2 can win if and only if at least $(\gamma+1)!+1$ key votes change their opinion. We let the threshold of every key voter be $\alpha = \binom{\gamma}{2}$, whereas $T = \binom{\gamma}{2}$.

Since by assumption there exists an FPT algorithm parameterized by k and T for the EMMI problem, we can use the algorithm to compute a minimal cost solution. This solution selects at least $(\gamma+1)!+1$ key votes from B, with every key voter connected to at least $\alpha = \binom{\gamma}{2}$ strategies in A by the threshold constraint. We check the total cost of this solution:

If G admits a γ-clique, then the total cost is at most $\alpha = \binom{\gamma}{2}$, since there always exist α vertices in A with at least $\lceil n^{\frac{\gamma}{\gamma+6}} \rceil > (k+1)!+1$ common neighbors.

If G does not admit a γ-clique, then the total cost is at least $\alpha+1 = \binom{\gamma}{2}+1$, since every α vertices in A have at most $(\gamma+1)!$ common neighbors, which means a solution with cost α cannot make any voter change his/her opinion.

The overall running time of the algorithm is $f(k,T)n^{O(1)} = f(\gamma)n^{O(1)}$ for some computable function f, which contradicts FPT $\neq$ W[1].

From the proof of Theorem 1, we can directly have the following corollary.

Corollary 1. *Even if there are only two candidates and all the strategies have the same cost, the EMMI problem is NP-hard.*

4 Algorithms for a Special Setting of EMMI

In the previous section, we have shown strong inapproximability results for the EMMI problem. It is not difficult to see that the hardness follows from the fact that the connections between strategies and voters are arbitrary, i.e., one strategy may influence an arbitrary subset of voters.

In reality, there are usually restrictions on how they are connected. Taking the TV channel as an example. The number of worldwide TV channels is rare. While nation or state-wide TV channels are much more common. Due to language or different favor of living habits, the influence of such local TV channels is limited to the physical regional. It is suitable to assume the influence set of each TV channel forms the laminar structure.

Definition 1. *If for arbitrary two strategies $s_h, s_k \in S$ at least one of the following three property holds, i) $I_h \supseteq I_k$; ii) $I_k \supseteq I_h$; iii) $I_k \cap I_h \neq \emptyset$, then we call the influence strategies forms a laminar structure.*

In this section, we consider one of the fundamental subclasses of the EMMI problem: EMMI-Laminar. It is very natural in real elections and meanwhile of significant importance in theory. Moreover, it is general enough to incorporate the classical bribery problem as a special case.

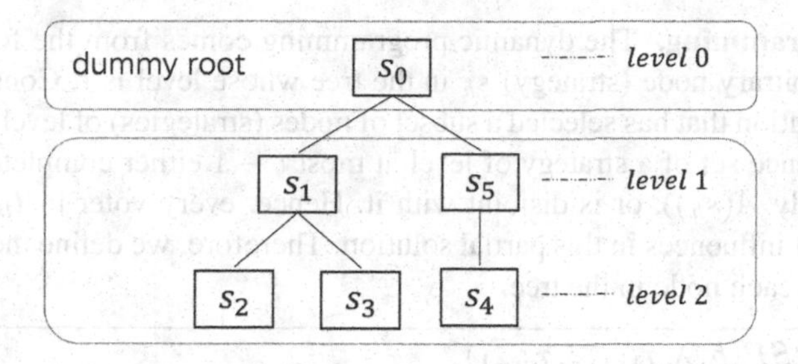

Fig. 2. A Tree Representation of the Instance Introduced in Sect. 2 (see Fig. 1)

4.1 A Dynamic Programming for EMMI-Laminar

If the influence strategy sets of the EMMI problem form the laminar structure, then we call it the EMMI-Laminar problem. The goal of this subsection is to prove the following theorem.

Theorem 3. *There exists an algorithm for the EMMI-Laminar problem that runs in* $O(r^3 n^{3m-3})$ *time.*

For convenience, we say strategy s_h dominates s_k if $I_k \subsetneq I_h$ or $I_k = I_h, h > k$. That is, in the Laminar Structure, any two strategies either cover two completely different subsets of voters, or one of them dominates the other. For example, in Fig. 1, strategy s_1 dominates strategy s_2 and s_3 and meanwhile no strategies dominates s_1.

A Forest/Tree Representation. We use a forest to represent a laminar structure. Each node in the forest represents a strategy (and its influence set). We define the *level* of node iteratively as follows: Initially, all nodes that are not dominated by any other nodes are labeled with level 1; Suppose we have labeled nodes up to level ℓ, then among all the unlabeled nodes, each one that is not dominated by any other unlabeled node is labeled as level $\ell + 1$. Repeat this procedure, until all nodes are labeled. Finally, we connect each level $\ell + 1$ node to every node of level ℓ that dominates it. By doing so, a forest is constructed.

For ease of presentation, we add a dummy node s_0 that connects all the level 1 strategies to make the forest into a tree (the influence set I_0 of the dummy strategy s_0 contains all the voters, and its cost $q_0 = B + 1$). Since, the cost of the dummy strategy s_0 is greater than the total budget of the attacker. Hence, any feasible solution would not select s_0. It is easy to see that these two instances are equivalent.

Now, we are able to talk about the EMMI problem on the tree representation. Selecting a node in the tree means adopting the corresponding strategy. In the following part, we will introduce the dynamic programming based on the tree representation.

We denote by $R(s_h)$ the set of strategies that are dominated by strategy s_h together with s_h itself, which is essentially the subtree rooted at s_h in the tree. We denote by $A(s_h)$ the set of strategies that dominate strategy h, which are essentially ancestors of s_h in the tree. For example, in Fig. 2, $A(s_1) = \{s_0\}$ and $R(s_1) = \{s_1, s_2, s_3\}$.

Dynamic Programming. The dynamic programming comes from the following fact. Consider an arbitrary node (strategy) s_h in the tree whose level is ℓ. Consider an arbitrary partial solution that has selected a subset of nodes (strategies) of level at most $\ell-1$. Since the influence set of a strategy of level at most $\ell - 1$ either completely covers I_h (which is exactly $A(s_h)$), or is disjoint with it. Hence, every voter in I_h receives the same number of influences in this partial solution. Therefore, we define the subproblem associated with each node in the tree.

Subproblem $SP(h, \lambda, \tau_1, \ldots, \tau_{m-1})$
For an arbitrary node s_h in the tree. Suppose exactly λ nodes are selected in $A(s_h)$. Select a subset of nodes $W \subseteq R(s_h)$ minimizing $\sum_{j\in W} q_j$ and satisfying there are at least τ_i voters in $I_h \cap V_i$ activated for $1 \le i \le m-1$.

We denote by $SP(h, \lambda, \tau_1, \tau_2, \cdots, \tau_{m-1})$ the above subproblem. Abusing the notations a bit, we also use $SP(h, \lambda, \tau_1, \tau_2, \cdots, \tau_{m-1})$ to denote the optimal objective value of this subproblem. It should be clear that since $\lambda \le |S| = r$ and $\tau_i \le |V_i| \le n$, the total number of different subproblems is bounded by $O(r^2 n^{m-1})$, which is polynomial if m is a constant.

The optimal objective values of subproblems can be calculated in a bottom-up way along the tree.

Initial Values. Initially, we can calculate $SP(h, \lambda, \tau_1, \tau_2, \cdots, \tau_{m-1})$ for every leaf node s_h in the tree: Consider all the voters in I_h. A voter is activated if the number of influences the voter receiving exceeds his/her threshold. Let $\phi_{h,i}(\lambda)$ be the total number of activated voters in $I_h \cap V_i$ when every voters in I_h receives influence λ times. Then we define

(i) $SP(h, \lambda, \tau_1, \tau_2, \cdots, \tau_{m-1}) = 0$, if $\phi_{h,i}(\lambda) \ge \tau_i$ for every $1 \le i \le m-1$;
(ii) $SP(h, \lambda, \tau_1, \tau_2, \cdots, \tau_{m-1}) = q_h$, if (i) does not hold, and $\phi_{h,i}(\lambda) + 1 \ge \tau_i$ for every $1 \le i \le m-1$;
(iii) $SP(h, \lambda, \tau_1, \tau_2, \cdots, \tau_{m-1}) = \infty$, if (i) and (ii) do not hold.

The above is straightforward: If (i) holds, then the answer of the subproblem for the leaf h is true without selecting h, whereas the optimal objective value is 0; If (i) does not hold, yet τ_i voters in $I_h \cap V_i$ can be activated by additionally selecting h, then the optimal objective value is the cost of h; Otherwise, the subproblem has no solution and we let the optimal objective value be ∞.

Recursive Calculation. The value of $SP(h, \lambda, \tau_1, \tau_2, \cdots, \tau_{m-1})$ for a non-leaf node s_h can be calculated based on the values from the children of h. Suppose the node s_h has u children which are $s_{h_1}, \ldots, s_{h_u}$. We define the following "sub-subproblem":

Sub-subproblem-1 $\Gamma_1(h, \lambda, \tau_1, \ldots, \tau_{m-1})$
For an arbitrary node s_h in the tree. Suppose exactly λ nodes are selected in $A(s_h)$. Select a subset of nodes $W \subseteq R(s_h) \setminus s_h$ minimizing $\sum_{j\in W} q_j$ and satisfying there are at least τ_i voters in $I_h \cap V_i$ activated for $1 \le i \le m-1$.

We have the following claims.

Claim. Abuse the notation a little bit. Let $\Gamma_1(h, \lambda, \tau_1, \ldots, \tau_{m-1})$ be the minimal objective of Sub-subproblem-1, then

$$\Gamma_1(h, \lambda, \tau_1, \ldots, \tau_{m-1}) = \min_{\tau_1^1, \cdots, \tau_{m-1}^1, \cdots, \tau_1^u, \cdots, \tau_{m-1}^u} \{\textstyle\sum_{j=1}^u SP(h_j, \lambda, \tau_1^j, \cdots, \tau_{m-1}^j) : \\ \textstyle\sum_{j=1}^u \tau_i^j = \tau_i - \chi_{h,i}(\lambda), 1 \le i \le m-1\} \quad (1)$$

where $\chi_{h,i}(\lambda)$ is the total number of voters in $\left(I_h \setminus \cup_{j=1}^u I_{h_j}\right) \cap V_i$ whose threshold is bounded by λ.

Proof. Since strategy s_h is not chosen, every voter in I_{h_j} also receives influences λ times from s_h together with the nodes of a smaller level. The τ_i activated voters within I_h either comes from some subset I_{h_j}, which is counted in τ_i^j, or comes from $I_h \setminus \cup_{j=1}^u I_{h_j}$, which is counted in $\chi_{h,i}(\lambda)$. Hence Eq (1) is true.

Similarly, we can define another closely related "sub-subproblem".

Sub-subproblem-2 $\Gamma_2(h, \lambda, \tau_1, \ldots, \tau_{m-1})$
For an arbitrary node s_h in the tree. Suppose s_h together with exactly λ nodes in $A(s_h)$ are selected. Select a subset of nodes $W \subseteq R(s_h) \setminus s_h$ minimizing $\sum_{j \in W} q_j$ and satisfying there are at least τ_i voters in $I_h \cap V_i$ activated for $1 \le i \le m-1$.

Claim. Abuse the notation a little bit. Let $\Gamma_2(h, \lambda, \tau_1, \ldots, \tau_{m-1})$ be the minimal objective of Sub-subproblem 2, then

$$\Gamma_2(h, \lambda, \tau_1, \ldots, \tau_{m-1}) = \min_{\tau_1^1, \cdots, \tau_{m-1}^1, \cdots, \tau_1^u, \cdots, \tau_{m-1}^u} \{\textstyle\sum_{j=1}^u SP(h_j, \lambda+1, \tau_1^j, \cdots, \tau_{m-1}^j) : \\ \textstyle\sum_{j=1}^u \tau_i^j = \tau_i - \chi_{h,i}(\lambda+1), 1 \le i \le m-1\}, \quad (2)$$

where $\chi_{h,i}(\lambda+1)$ is the total number of voters in $\left(I_h \setminus \cup_{j=1}^u I_{h_j}\right) \cap V_i$ whose threshold is bounded by $\lambda+1$.

Proof. Since strategy s_h is chosen, every voter in I_h (and also every I_{h_j}) not only receives influence λ times from nodes of a smaller level than h, but also receives influence once from s_h, and thus receives influences $\lambda+1$ times in total from s_h together with nodes of a smaller level than h. The τ_i activated voters within I_h either comes from some subset I_{h_j}, which is counted in τ_i^j, or comes from $I_h \setminus \cup_{j=1}^u I_{h_j}$, which is counted in $\chi_{h,i}(\lambda+1)$.

Given the above two claims, it is now clear that

$$SP(h, \lambda, \tau_1, \tau_2, \cdots, \tau_{m-1}) = \min\{\Gamma_1(h, \lambda, \tau_1, \ldots, \tau_{m-1}), \Gamma_2(h, \lambda, \tau_1, \ldots, \tau_{m-1}) + q_h\}.$$

We showed above the correctness of Eq. (1) and (2). However, directly using Eq. (1), (2) to calculate Γ_1, Γ_2 need $n^{O(nm)}$ time which is unaffordable. Observing that Eq. (1), (2) are very similar. Hence, we adopt the following recursive function f to calculate Γ_1, Γ_2 simultaneously.

$$f(k, \lambda, \tau_1, \cdots, \tau_{m-1}) = \begin{cases} SP(h_1, \lambda, \tau_1, \cdots, \tau_{m-1}), & \text{if } k = 1 \\ \min_{x_1 \in [0,\tau_1], \cdots, x_{m-1} \in [0, \tau_{m-1}]} \{f(k-1, \lambda, x_1, \cdots, x_{m-1}) \\ \quad + SP(h_k, \lambda, \tau_1 - x_1, \cdots, \tau_{m-1} - x_{m-1})\} & \text{if } 1 < k \leq u \end{cases}$$

We know that

$$\Gamma_1(h, \lambda, \tau_1, \ldots, \tau_{m-1}) = f(u, \lambda, \tau_1 - \chi_{h,1}(\lambda), \cdots, \tau_{m-1} - \chi_{h,m-1}(\lambda))$$

and

$$\Gamma_2(h, \lambda, \tau_1, \ldots, \tau_{m-1}) = f(u, \lambda, \tau_1 - \chi_{h,1}(\lambda + 1), \cdots, \tau_{m-1} - \chi_{h,m-1}(\lambda + 1)).$$

Lemma 1. *The optimal solution of each subproblem $SP(h, \lambda, \tau_1, \ldots, \tau_{m-1})$ can be calculated in $O(rn^{2m-2})$ time.*

Proof. Once $\Gamma_1(h, \lambda, \tau_1, \ldots, \tau_{m-1})$ and $\Gamma_1(h, \lambda, \tau_1, \ldots, \tau_{m-1})$ are known, the optimal solution of the subproblem can be calculated within $O(1)$ time. In order to calculate Γ_1, Γ_2, we need to calculate the value of the recursive function f for at most $O(rn^{m-1})$ times. And calculating the value of f once cost $O(n^{m-1})$ time.

After we calculate the values of all possible subproblems. We check subproblems associated with the dummy root s_0. For each $(\tau_1, \cdots, \tau_{m-1})$, if bribing τ_i voters from each V_i can make the designated candidate wins, then $(\tau_1, \cdots, \tau_{m-1})$ is called a feasible vector. Let $\mathscr{F}$ be the set of all feasible vectors, then the minimum bribing cost B_{opt} to let the designated candidate wins is given by the following:

$$B_{opt} = \min_{(\tau_1, \cdots, \tau_{m-1}) \in \mathscr{F}} \{SP(0, 0, \tau_1, \cdots, \tau_{m-1})\}.$$

If $B_{opt} \leq B$, then the answer of the EMMI-Laminar problem is "yes". Otherwise, the answer is "no".

Above all, we need to solve $O(r^2 n^{m-1})$ different subproblems. Solving each subproblem cost $O(rn^{2m-2})$ time. Hence, the overall running time of the dynamic programming is $O(r^3 n^{3m-3})$ and Theorem 3 is true. It is easy to see that if the number of candidates m is a constant, then it is a polynomial time algorithm for the EMMI-Laminar problem.

5 Conclusion

We introduced the EMMI problem as a realistic variant of the classic election manipulation problem. The EMMI problem models how attackers, which may or may not be the election candidates, attempt to exploit media advertisements to influence voters. We prove the hardness of the EMMI problem in the general setting and present a polynomial time algorithm for a special case of the EMMI problem, by leveraging a specific voter-strategy structure.

Acknowledgements. This material is based upon work supported by the "New Generation of AI 2030" major project (2018AAA0100902) and the National Science Foundation under grant no. 1433817.

References

1. Brandt, F., Conitzer, V., Endriss, U., Lang, J., Procaccia, A.D.: Handbook of Computational Social Choice. Cambridge University Press, New York (2016)
2. Bredereck, R., Chen, J., Faliszewski, P., Nichterlein, A., Niedermeier, R.: Prices matter for the parameterized complexity of shift bribery. Inf. Comput. **251**, 140–164 (2016)
3. Bredereck, R., Faliszewski, P., Niedermeier, R., Talmon, N.: Complexity of shift bribery in committee elections. In: Proceedings of the 30th AAAI Conference on Artificial Intelligence, pp. 2452–2458 (2016)
4. Brelsford, E., Faliszewski, P., Hemaspaandra, E., Schnoor, H., Schnoor, I.: Approximability of manipulating elections. In: Proceedings of the 23rd AAAI Conference on Artificial Intelligence, vol. 1, pp. 44–49 (2008)
5. Castiglioni, M., Ferraioli, D., Gatti, N.: Election control in social networks via edge addition or removal. In: Proceedings of the 34th AAAI Conference on Artificial Intelligence, vol. 34, pp. 1878–1885 (2020)
6. Chekuri, C., Clarkson, K.L., Har-Peled, S.: On the set multicover problem in geometric settings. ACM Trans. Algorithms **9**(1), 1–17 (2012)
7. Chen, L., et al.: Protecting election from bribery: new approach and computational complexity characterization. In: Proceedings of the 17th International Conference on Autonomous Agents and Multiagent Systems, pp. 1894–1896 (2018)
8. Corò, F., Cruciani, E., D'Angelo, G., Ponziani, S.: Vote for me! Election control via social influence in arbitrary scoring rule voting systems. In: Proceedings of the 18th International Conference on Autonomous Agents and Multiagent Systems, pp. 1895–1897 (2019)
9. Elkind, E., Faliszewski, P.: Approximation algorithms for campaign management. In: Internet and Network Economics, pp. 473–482 (2010)
10. Elkind, E., Faliszewski, P., Slinko, A.: Swap bribery. In: Proceedings of the 2nd International Symposium on Algorithmic Game Theory, pp. 299–310 (2009)
11. Faliszewski, P., Gonen, R., Koutecký, M., Talmon, N.: Opinion diffusion and campaigning on society graphs. In: Proceedings of the 27th International Joint Conference on Artificial Intelligence, pp. 219–225 (2018)
12. Faliszewski, P., Hemaspaandra, E., Hemaspaandra, L.A.: How hard is bribery in elections? J. Artif. Intell. Res. **35**, 485–532 (2009)
13. Faliszewski, P., Hemaspaandra, E., Hemaspaandra, L.A.: Multimode control attacks on elections. J. Artif. Intell. Res. **40**, 305–351 (2011)
14. Faliszewski, P., Reisch, Y., Rothe, J., Schend, L.: Complexity of manipulation, bribery, and campaign management in bucklin and fallback voting. In: Proceedings of the 14th International Conference on Autonomous Agents and Multiagent Systems, pp. 1357–1358 (2014)
15. Fowler, E.F., Franz, M., Ridout, T.N.: The big lessons of political advertising in 2018. The Conversation (2018). http://theconversation.com/the-big-lessons-of-political-advertising-in-2018-107673. Accessed 3 Dec
16. Guo, J., Niedermeier, R.: Exact algorithms and applications for tree-like weighted set cover. J. Discrete Algorithms **4**(4), 608–622 (2006)
17. Hua, Q.S., Yu, D., Lau, F.C., Wang, Y.: Exact algorithms for set multicover and multiset multicover problems. In: Proceedings of the 20th International Symposium on Algorithms and Computation, pp. 34–44 (2009)
18. Lin, B.: The parameterized complexity of k-biclique. In: Proceedings of the 26th Annual ACM-SIAM Symposium on Discrete Algorithms, pp. 605–615 (2014)
19. Mehrizi, M.A., D'Angelo, G.: Multi-winner election control via social influence: hardness and algorithms for restricted cases. Algorithms **13**(10), 251 (2020)

20. Parkes, D.C., Xia, L.: A complexity-of-strategic-behavior comparison between Schulze's rule and ranked pairs. In: Proceedings of the 26th AAAI Conference on Artificial Intelligence, pp. 1429–1435 (2012)
21. Van Bevern, R., Chen, J., Hüffner, F., Kratsch, S., Talmon, N., Woeginger, G.J.: Approximability and parameterized complexity of multicover by c-intervals. Inf. Process. Lett. **115**(10), 744–749 (2015)
22. Wilder, B., Vorobeychik, Y.: Controlling elections through social influence. In: Proceedings of the 17th International Conference on Autonomous Agents and Multiagent Systems, pp. 265–273 (2018)
23. Xia, L.: Computing the margin of victory for various voting rules. In: Proceedings of the 13th ACM Conference on Electronic Commerce, pp. 982–999 (2012)

Improved Approximation Algorithms for Multiprocessor Scheduling with Testing

Mingyang Gong and Guohui Lin(✉)

Department of Computing Science, University of Alberta,
Edmonton, AB T6G 2E8, Canada
{mgong4,guohui}@ualberta.ca

Abstract. Multiprocessor scheduling, also called scheduling on parallel identical machines to minimize the makespan, is a classic optimization problem that has received numerous studies. Scheduling with testing is an online variant where the processing time of a job is revealed by an extra test operation, or otherwise the job has to be executed for a given upper bound on the processing time. Albers and Eckl recently studied the multiprocessor scheduling with testing; among others, for the non-preemptive setting they presented an approximation algorithm with competitive ratio approaching 3.1016 when the number of machines tends to infinity and an improved approximation algorithm with competitive ratio approaching 3 when all test operations take a time unit. We propose to first sort the jobs into the non-increasing order of the minimum value between the upper bound and the testing time, then partition the jobs into three groups and process them group by group according to the sorted job order. We show that our algorithm achieves improved competitive ratios, which approach 2.9513 when the number of machines tends to infinity in the general case; when all test operations take a time unit, our algorithm achieves even better competitive ratios approaching 2.8081.

Keywords: Multiprocessor scheduling · scheduling with testing · makespan · non-preemptive · competitive ratio · approximation algorithm

1 Introduction

We consider the multiprocessor scheduling with testing to minimize the makespan, and we continue using the notations introduced in [2]. In this online scheduling problem, a set of jobs $\mathcal{J} = \{J_1, J_2, \ldots, J_n\}$ are given to be processed on several parallel identical machines $\mathcal{M} = \{M_1, M_2, \ldots, M_m\}$. Each job J_j is given with an upper bound u_j on its processing time and a testing time t_j if one wishes to reveal the accurate processing time p_j. The two parameters u_j and t_j are given in advance so that the scheduler can take full advantage of, but

J. Chen et al. (Eds.): IJTCS-FAW 2021, LNCS 12874, pp. 65–77, 2022.
https://doi.org/10.1007/978-3-030-97099-4_5

the accurate processing time p_j with $0 \leq p_j \leq u_j$ is revealed only after the test operation is executed. That is, an algorithm has to decide whether to test the job J_j or not when processing J_j; and if J_j is untested, then the processing time of J_j in the algorithm is u_j, or otherwise by spending t_j time units the algorithm knows of the actual processing time being p_j.

One clearly sees that for any job J_j, if $t_j > u_j$, then an algorithm would not choose to execute the test operation for the job J_j at all. In our approximation algorithms below, for most jobs, we choose not to execute the test operation if $T(m)t_j > u_j$, where the threshold $T(m) \geq (1+\sqrt{5})/2 \approx 1.6180$ is a function depending on the number m of machines we have. We denote $\varphi = (1+\sqrt{5})/2$, the golden ratio.

Let $\tau_j = \min\{t_j, u_j\}$ and $\rho_j = \min\{p_j + t_j, u_j\}$, $j = 1, 2, \ldots, n$, the latter of which is the processing time of the job J_j in an offline algorithm. Given a schedule, if J_j is scheduled on M_i, then the *completion time* C_j of J_j is the total processing time of the jobs scheduled before J_j on M_i plus the processing time of J_j. The *load* of M_i is the total processing time of all the jobs scheduled on it. Note that our goal is to minimize the makespan, which is the largest completion time among all jobs. In this paper we consider only the non-preemptive setting, where testing or processing of a job on a machine is not to be interrupted. In the three-field notation, this online problem with the number m of machines being part of the input is denoted as $P \mid t_j, 0 \leq p_j \leq u_j \mid C_{\max}$.

For an online algorithm A, let $C^A(I)$ denote the makespan of the schedule produced by A for the instance I; let $C^*(I)$ denote the optimal offline makespan for the instance I. The performance of the algorithm A is measured by the competitive ratio $\sup_I\{C^A(I)/C^*(I)\}$.

Albers and Eckl [2] have summarized excellently the related work on multiprocessor scheduling (where all job processing times are given ahead), online multiprocessor scheduling (where the jobs arrive in sequence and the scheduler needs to assign a job at its arrival), and semi-online multiprocessor scheduling (where some piece of information except the job processing times is available to the scheduler). The interested readers are referred to [2] for the details and references therein. Besides, for online multiprocessor scheduling, it is known that the list scheduling algorithm is optimal when $m = 2, 3$ [6,8]. When $m = 4$, it admits a 26/15-competitive algorithm while the best competitive ratio lower bound is $\sqrt{3}$ [12]. Lower bounds were also given for several small values of m in [4], and the optimality of existing online algorithms for $m = 4, 5, 6$ are proven based on pseudo lower bounds in [13]. As also mentioned in [2], the current best 1.9201-competitive algorithm is due to Fleischer and Wahl [7], and the current best lower bound of 1.88 is due to Rudin [11].

In one specific semi-online multiprocessor scheduling problem, the scheduler is given in advance the maximum job processing time. For this problem, the algorithm by Fleischer and Wahl [7] is also the current best, while it is unknown whether this maximum job processing time is useful or not for designing better algorithms. Nevertheless, for some small values of m, better results do exist, including an optimal 4/3-competitive algorithm when $m = 2$ [9], a competitive ratio lower bound of $\sqrt{2}$ [3] and a 3/2-competitive algorithm when $m = 3$ [14], and a competitive ratio lower bound of $(3+\sqrt{33})/6$ when $m \geq 4$ and a $(\sqrt{5}+1)/2$-competitive algorithm when $m = 4$ [10].

For scheduling with testing, one explores the uncertainty, which is the accurate job processing time in our case, through queries with a given cost, which is the test operation that requires a certain amount of time. For general uncertainty exploration, the readers again may refer to [2] for more details and references therein, such as the optimization problems that have been studied in this setting and what information can be explored. Specific to scheduling with testing, Dürr et al. [5] first investigated the single machine scheduling to minimize the total job completion time or to minimize the makespan, where all test operations need one unit of time, called the *uniform testing case*. Albers and Eckl [1] considered the general testing times (called the *general testing case*) and presented generalized algorithms for both objectives. For scheduling with testing on parallel identical machines to minimize the makespan, that is, $P \mid t_j, 0 \leq p_j \leq u_j \mid C_{\max}$ and $P \mid t_j = 1, 0 \leq p_j \leq u_j \mid C_{\max}$, Albers and Eckl [2] considered both non-preemptive and preemptive settings. For the preemptive setting, the authors presented a 2-competitive algorithm for either problem, which is asymptotically optimal. For the non-preemptive setting, they presented the SBS algorithm (in which the jobs are partitioned into three groups S_1, B, S_2 and processed in order) for $P \mid t_j, 0 \leq p_j \leq u_j \mid C_{\max}$ and showed that its competitive ratio is at most 3.1016; they also presented the 3-competitive U-SBS algorithm (here U stands for uniform) for $P \mid t_j = 1, 0 \leq p_j \leq u_j \mid C_{\max}$.

In this paper, we focus on the non-preemptive multiprocessor scheduling with testing problems, the general case $P \mid t_j, 0 \leq p_j \leq u_j \mid C_{\max}$ and the uniform case $P \mid t_j = 1, 0 \leq p_j \leq u_j \mid C_{\max}$, and aim to design better approximation algorithms. We adopt the idea in the algorithms by Albers and Eckl [2] to set up a threshold $T(m)$, where m is the number of machines, so that a job J_j with $T(m)t_j > u_j$ is tested; the new ingredient is to sort the jobs into the non-increasing order of their τ-values, partition them into three groups, and then process them group by group according to the sorted job order. In other words, we do not use an arbitrary job order as in the algorithms by Albers and Eckl [2], but take full advantage of the known t-values and u-values to sort the jobs into a specific order. We show that such a job order enables us to set up a better threshold $T(m)$, which eventually gives rise to better competitive ratios. At the end, we achieve an algorithm for the general case $P \mid t_j, 0 \leq p_j \leq u_j \mid C_{\max}$, with its competitive ratio approaching $\varphi + \frac{4}{3} \approx 2.9513$ when the number m of machines tends to infinity, and an algorithm for the uniform case $P \mid t_j = 1, 0 \leq p_j \leq u_j \mid C_{\max}$, with its competitive ratio approaching $\frac{7+\sqrt{97}}{6} \approx 2.8081$. For some values of m, we list the competitive ratios of the SBS algorithm, the U-SBS algorithm, and our two algorithms called BBS and U-BBS, in Table 1.

The rest of the paper is organized as follows: In Sect. 2, we give the definitions of the threshold function $T(m)$ in the number m of machines, for the general case and the uniform case; we also characterize some basic properties of the function.

Table 1. The achieved (in bold) and the previous best competitive ratios when there are m parallel identical machines

m	1	2	3	4	5	10	100	∞
SBS [2]	1.6180	2.3806	2.6235	2.7439	2.8158	2.9591	3.0874	3.1016
BBS	1.6180	**2.3133**	**2.5069**	**2.6180**	**2.6847**	**2.8180**	**2.9380**	**2.9513**
U-SBS [2]	1.6180	2.3112	2.5412	2.6560	2.7248	2.8625	2.9862	3
U-BBS	1.6180	**2.2707**	**2.3985**	**2.5000**	**2.5612**	**2.6843**	**2.7957**	**2.8081**

In Sect. 3 and Sect. 4, we present our BBS algorithm and U-BBS algorithm, respectively, and analyze their competitive ratios. Section 5 concludes the paper.

2 Preliminaries

In an instance of the $P \mid t_j, 0 \le p_j \le u_j \mid C_{\max}$ problem, we are given a set of n jobs to be processed on m parallel identical machines, where for each job J_j, its t_j and u_j are known but the value of p_j is revealed only after testing. Recall that $\tau_j = \min\{t_j, u_j\}$. We first sort the jobs into the non-increasing order of the τ-values in $O(n \log n)$ time, and assume without loss of generality in the sequel that the (ordered) job set is $\mathcal{J} = \{J_1, J_2, \ldots, J_n\}$ satisfying $\tau_1 \ge \tau_2 \ge \ldots \ge \tau_n$ and if $\tau_i = \tau_j$, then $i < j$ implies $u_i \ge u_j$. Note that $\varphi = \frac{\sqrt{5}+1}{2}$ is the golden ratio. We define the threshold functions in the number m of machines (the variable m as the subscript) as

$$T_m^g = \begin{cases} \frac{3\varphi+6+\sqrt{45\varphi+213}}{14}, & m = 2, \\ \frac{3m\varphi+4m-4}{4m-1}, & m = 1 \text{ or } m \ge 3 \end{cases} \tag{1}$$

and

$$T_m^u = \begin{cases} \frac{9+3\sqrt{37}}{14}, & m = 2, \\ \frac{7m-4+\sqrt{97m^2--68m+16}}{2(4m-1)}, & m = 1 \text{ or } m \ge 3 \end{cases} \tag{2}$$

for the general case and the uniform case, respectively. Note that we use superscripts 'g' and 'u' to denote the two cases.

One sees that $T_1^g = \varphi = T_1^u$. We characterize the properties of T_m^g and T_m^u in the next two lemmas, respectively. Due to space limit, we skip the proofs of them and some other lemmas below.

Lemma 1. *(i) If $m \ge 2$, then $1 + \frac{1}{T_m^g} \le \varphi \le T_m^g$.*

(ii) $\varphi + 1 - \frac{1}{m} \le \frac{(4m-1)T_m^g}{3m}$.
(iii) If $m \ge 4$, then $1 + \frac{1}{3\varphi} - \frac{4}{3m\varphi} - \frac{1}{T_m^g \varphi} > \frac{2}{3}$.
(iv) $1 - \frac{1}{9\varphi} - \frac{1}{T_3^g \varphi} > \frac{3}{5}$.
(v) If $m \ge 2$, then $\frac{(5m-2)T_m^g}{6m\varphi} > \frac{2}{3}$.

Lemma 2. *(i) If* $m \geq 2$*, then* $1 + \frac{1}{T_m^u} \leq \varphi \leq T_m^u$.

(ii) $2 + \frac{1}{T_m^u} - \frac{1}{m} \leq \frac{(4m-1)T_m^u}{3m}$.

(iii) If $m \geq 4$*, then* $T_m^u \geq 2$ *and* $\frac{2}{m} + \frac{1}{T_m^u} \leq 1$.

(iv) $\frac{2}{3}(2 + \frac{2}{T_3^u}) \leq \frac{17}{9} + \frac{1}{T_3^u}$.

(v) If $m \geq 2$*, then* $\frac{(5m-2)T_m^u}{6m\varphi} > \frac{2}{3}$.

We next partition the jobs into three groups B_1, B_2^a, S^a using the threshold function T_m^a, where $a = g, u$. For each group, the jobs inside are ordered, inherited from the original non-increasing order of τ-values, that is, the job order for a group is a subsequence of the original job sequence $\langle J_1, J_2, \ldots, J_n \rangle$. Our algorithm processes first the jobs of B_1 in their order, then the jobs of B_2^a in their order, and lastly the jobs of S^a in their order, and it assigns each job to the least loaded machine (tie broken to the smallest index); we call it the BBS algorithm in the general testing case and the U-BBS algorithm in the uniform testing case.

If $n \leq m$, then $B_1 = \mathcal{J}$, $B_2^a = S^a = \emptyset$.

If $n > m$, then

$$\begin{aligned} B_1 &= \{J_1, J_2, \ldots, J_m\}, \\ B_2^a &= \left\{ J_j \in \mathcal{J} \setminus B_1 : \frac{u_j}{t_j} \geq T_m^a \right\}, \\ S^a &= \mathcal{J} \setminus (B_1 \cup B_2^a). \end{aligned} \tag{3}$$

That is, the jobs of B_1 have the largest τ-values, and the jobs of B_2^a have the ratios larger than the threshold so that they will be tested to obtain the accurate processing times.

Recall that $\rho_j = \min\{t_j + p_j, u_j\}$ is the processing time of the job J_j in an offline algorithm. Let p_j^B denote the processing time of J_j in our algorithm, which is $t_j + p_j$ if the job is tested or otherwise u_j. We define $\alpha_j = \frac{p_j^B}{\rho_j}$, which is greater than or equal to 1.

Let C^B and C^* be the makespan of the schedule by our algorithm and the offline optimal makespan for the instance, respectively. For performance analysis, we may assume w.l.o.g. that the makespan C^B is determined by the job J_n. The following lower bounds on C^* are easy to see:

$$C^* \geq \left\{ \max_{j=1}^{n} \rho_j, \frac{1}{m} \sum_{j=1}^{n} \rho_j \right\}. \tag{4}$$

In the schedule by our algorithm, we assume there are ℓ machines each is assigned with only one job of B_1 before executing the job J_n, and let $\mathcal{J}^a$ denote the subset of these jobs of B_1, for $a = g, u$. According to our algorithm, we can assume that $\mathcal{J}^a = \{J_{i_1}, J_{i_2}, \ldots, J_{i_\ell}\} \subseteq B_1$ and J_{i_j} is scheduled on M_{i_j}, $j = 1, 2, \ldots, \ell$. We note that if $\ell = 0$, then $\mathcal{J}^a = \emptyset$.

3 The General Testing Case

With the threshold function T_m^g defined in Eq. (1) and the job groups B_1, B_2^g, S^g formed by Eq. (3), the BBS algorithm for the $P \mid t_j, 0 \leq p_j \leq u_j \mid C_{\max}$ problem can be presented as the following Algorithm 1. We remind the readers that the jobs are given in the non-increasing order of τ-values, with an overhead of $O(n \log n)$ time.

Algorithm 1. BBS Algorithm

Input: B_1, B_2^g, S^g;
for $J_j \in B_1$ in increasing order of j **do**
  if $\frac{u_j}{t_j} < \varphi$ **then**
    schedule J_j untested on machine M_j;
  else
    test and schedule J_j on machine M_j;
  end if
end for
for $J_j \in B_2^g$ in increasing order of j **do**
  test and schedule J_j on the least loaded machine;
end for
for $J_j \in S^g$ in increasing order of j **do**
  schedule J_j untested on the least loaded machine;
end for

Lemma 3. *For the job J_j, $j = 1, 2, \ldots, n$, we have*

$$\alpha_j \leq \begin{cases} \varphi, & J_j \in B_1, \\ 1 + \frac{1}{T_m^g}, & J_j \in B_2^g, \\ T_m^g, & J_j \in S^g. \end{cases}$$

Lemma 4. *If $J_n \in B_2^g$ and $m \geq 2$, then we have $C^B \leq \frac{(4m-1)T_m^g}{3m} C^*$.*

Proof. Since $J_n \in B_2^g$, we have $\frac{u_n}{t_n} \geq T_m^g$. So $\tau_n = t_n$ and by Lemma 3, we have $p_n^B \leq (1 + \frac{1}{T_m^g}) u_n$. Because J_n is scheduled on the least-loaded machine, by Eq. (4) Lemma 1(i) and Lemma 3, we obtain

$$\begin{aligned} C^B &\leq \frac{1}{m} \sum_{J_k \in \mathcal{J} \setminus J_n} p_k^B + p_n^B \leq \frac{1}{m} \sum_{J_k \in \mathcal{J}} p_k^B + \left(1 - \frac{1}{m}\right) p_n^B \\ &= \frac{\varphi}{m} \sum_{J_k \in \mathcal{J}} \rho_k + \left(1 - \frac{1}{m}\right) p_n^B \\ &\leq \varphi C^* + \left(1 - \frac{1}{m}\right) p_n^B. \end{aligned}$$

If $p_n^B = p_n + t_n \leq C^*$, then by Lemma 1(ii), we have

$$C^B \leq \left(\varphi + 1 - \frac{1}{m}\right) C^* \leq \frac{(4m-1)T_m^g}{3m} C^*.$$

Therefore we can assume $p_n + t_n > C^*$. It means $\rho_n = u_n \leq C^*$. Note that $\rho_j \geq \tau_j \geq \tau_n = t_n$, $\forall J_j \in B_1 \cup B_2^g$. So in the optimal solution, J_n must be scheduled separately on a machine; and $\mathcal{J} \setminus J_n$ are assigned on other $m-1$ machines. We distinguish two cases based on the value of m.

Case 1: $m \geq 3$. In this case, $\varphi + \frac{4}{3}(1 - \frac{1}{m}) = \frac{(4m-1)T_m^g}{3m}$. So it is sufficient to show the upper bound is $\varphi + \frac{4}{3}(1 - \frac{1}{m})$.

Subcase 1.1: $t_n \leq \frac{1}{3}C^*$. So we have

$$C^B \leq \varphi C^* + \left(1 - \frac{1}{m}\right)(t_n + u_n) \leq \left(\varphi + \frac{4}{3}\left(1 - \frac{1}{m}\right)\right) C^*.$$

Subcase 1.2: $t_n > \frac{1}{3}C^*$. Note that $\rho_j \geq \tau_j \geq \tau_n = t_n > \frac{1}{3}C^*$, $\forall J_j \in B_1 \cup B_2^g$. If $|B_2^g| \geq m$, then $|\mathcal{J} \setminus J_n| \geq 2m - 1$. Therefore in the optimal solution, there are at least 3 jobs in $\mathcal{J} \setminus J_n$ that will be assigned to the same machine, a contradiction. So we can assume $|B_2^g| \leq m - 1$. If there is a machine whose load is no more than $(\varphi + \frac{1}{3} - \frac{4}{3m} - \frac{1}{T_m^g})C^*$ before J_n arrives, then by Eq. (4) and Lemma 3, we have

$$C^B \leq \left(\varphi + \frac{1}{3} - \frac{4}{3m} - \frac{1}{T_m^g}\right) C^* + \left(1 + \frac{1}{T_m^g}\right) C^* = \left(\varphi + \frac{4}{3}\left(1 - \frac{1}{m}\right)\right) C^*.$$

So we can assume the loads of all the machines are at least $(\varphi + \frac{1}{3} - \frac{4}{3m} - \frac{1}{T_m^g})C^*$ before J_n comes.

(i) $m \geq 4$. By Lemma 1(iii) and Lemma 3, we have $\rho_{i_j} \geq (1 + \frac{1}{3\varphi} - \frac{4}{3m\varphi} - \frac{1}{T_m^g \varphi})C^* > \frac{2}{3}C^*$, $j = 1, 2, \ldots, \ell$. So in the optimal schedule, all jobs of $\mathcal{J}^g$ and J_n must be scheduled on the $\ell + 1$ machines, respectively. Note that there are $m - \ell - 1$ machines and at least $2(m - \ell)$ jobs left. So there is at least one machine with 3 jobs, a contradiction.

(ii) $m = 3$. Since $|B_2^g| \leq 2$, we know $|\mathcal{J}^g| = \ell \geq 2$.

By Lemma 1(iv), we have $\rho_{i_j} \geq (1 - \frac{1}{9\varphi} - \frac{1}{T_3^g \varphi})C^* > \frac{3}{5}C^*$, $j = 1, 2, \ldots, \ell$. If $\ell = 3$, then there are 4 jobs whose processing time is larger than $\frac{3}{5}C^*$ in the optimal solution, a contradiction. Otherwise, $\ell = 2$. So $|B_2^g| = 2$. We set $J_h = B_2^g \setminus J_n$. If $\rho_h \geq \frac{2}{5}C^*$, then we have three jobs with processing time at least $\frac{3}{5}C^*, \frac{3}{5}C^*, \frac{2}{5}C^*$ in the optimal schedule, respectively, a contradiction. Hence we can assume $\rho_h < \frac{2}{5}C^*$. If $\rho_j \geq \frac{2}{3}C^*$, $\forall J_j \in \mathcal{J}^g$, then there are two jobs whose processing times are larger than $\frac{2}{3}C^*$ in the optimal schedule. And $\rho_h \geq t_h > \frac{1}{3}C^*$, a contradiction. Therefore there is at least one job in $\mathcal{J}^g$ such that its processing time is less than $\frac{2}{3}C^*$. So by Lemma 3, we have

$$C^B \leq \frac{2\varphi}{3}C^* + t_n + u_n \leq \frac{2\varphi}{3}C^* + \rho_h + u_n \leq \left(\frac{2\varphi}{3} + \frac{7}{5}\right) C^*.$$

Case 2: $m = 2$. Recall that in the optimal solution, J_n must be scheduled separately on a machine. So we have $C^* \geq \sum_{i=1}^{n-1} \rho_i$ and by Lemma 3, $\sum_{i=1}^{n-1} p_i^B \leq \varphi C^*$. Since J_n is scheduled on the least-loaded machine, we obtain

$$C^B \leq \frac{\varphi}{2} C^* + \left(1 + \frac{1}{T_2^g}\right) C^* = \frac{7T_2^g}{6} C^*.$$

This proves the lemma. □

Lemma 5. *If $J_n \in S^g$ and $m \geq 2$, then we have $C^B \leq \frac{(4m-1)T_m^g}{3m} C^*$.*

Proof. Note that $J_n \in S^g$. Therefore $|B_1| = m$ and $p_n^B = u_n < T_m^g t_n$. If $\tau_n = u_n$, then $\rho_j \geq \tau_j \geq \tau_n = u_n = p_n^B$, $\forall J_j \in B_1 \cup S^g$. Otherwise, $\tau_n = t_n$ and then $\rho_j \geq \tau_j \geq \tau_n = t_n > \frac{p_n^B}{T_m^g}$, $\forall J_j \in B_1 \cup J_n$. Note that there are at least $m+1$ jobs in $B_1 \cup J_n$. So in the optimal solution, there are at least 2 jobs in $B_1 \cup J_n$ that will be assigned on the same machine. Hence

$$C^* > \frac{2p_n^B}{T_m^g}. \tag{5}$$

Case 1: $u_n \leq \frac{T_m^g}{3} C^*$. Because J_n is scheduled on the least-loaded machine, by Eq. (4), Lemma 1(i), we have

$$\begin{aligned}
C^B &\leq \frac{1}{m} \sum_{J_k \in \mathcal{J} \setminus J_n} p_k^B + p_n^B \leq \frac{1}{m} \sum_{J_k \in \mathcal{J}} p_k^B + \left(1 - \frac{1}{m}\right) p_n^B \\
&\leq T_m^g C^* + \left(1 - \frac{1}{m}\right) p_n^B \\
&= T_m^g C^* + \left(1 - \frac{1}{m}\right) u_n \\
&\leq \frac{(4m-1)T_m^g}{3m} C^*.
\end{aligned}$$

Case 2: $u_n > \frac{T_m^g}{3} C^*$.

Subcase 2.1: $|S^g| \geq m+1$. If $\tau_n = u_n$, then $\tau_n > \frac{T_m^g}{3} C^*$. Otherwise, $\tau_n = t_n > \frac{u_n}{T_m^g} > \frac{1}{3} C^*$. Therefore $\tau_n > \frac{1}{3} C^*$. Note that the BBS algorithm schedules all jobs in S^g in the non-increasing order of τ-values. So $\rho_j \geq \tau_j > \frac{1}{3} C^*$, $J_j \in B_1 \cup S^g$. Note that there are at least $2m+1$ jobs in $\mathcal{J}$. So in the optimal solution, at least 3 jobs will be assigned to the same machine, a contradiction.

Subcase 2.2: $|S^g| \leq m$.

If there exists a machine whose load is no more than $\frac{(5m-2)T_m^g}{6m} C^*$, then by Eq. (5), we have

$$C^B \leq \frac{(5m-2)T_m^g}{6m} C^* + \frac{T_m^g}{2} C^* = \frac{(4m-1)T_m^g}{3m} C^*.$$

So we can assume the loads of all machines are at least $\frac{(5m-2)T_m^g}{6m}C^*$ before J_n comes. Note that $\ell \geq m - |S^g| + 1$. By Lemma 1(v) and Lemma 3, we have $\rho_{i_j} \geq \frac{(5m-2)T_m^g}{6m\varphi}C^* > \frac{2}{3}C^*$, $j = 1, 2, \ldots, \ell$. Therefore in the optimal schedule, every job in $\mathcal{J}^g$ will be processed separately on a machine. So there are $|S^g| - 1$ machines and at least $2|S^g| - 1$ jobs left, suggesting one machine to process at least 3 jobs, a contradiction. □

Theorem 1. *When $m \geq 2$, the competitive ratio of the BBS algorithm is at most $\frac{(4m-1)T_m^g}{3m}$, which is increasing in m and approaches $\varphi + \frac{4}{3} \approx 2.9513$.*

Proof. Case 1: $J_n \in B_1$. By Eq. (4) and Lemma 3, we have $C^B = p_n^B \leq \varphi \rho_n \leq \varphi C^*$.

Case 2: $J_n \in B_2^g$. By Lemma 4, we have $C^B \leq \frac{(4m-1)T_m^g}{3m}C^*$.

Case 3: $J_n \in S^g$. By Lemma 5, we have $C^B \leq \frac{(4m-1)T_m^g}{3m}C^*$. This proves the competitive ratio. □

4 The Uniform Testing Case

Since $t_j = 1$, for all $j = 1, 2, \ldots, n$, we may assume that the jobs are given in the non-increasing order of u-values, that is, $u_1 \geq u_2 \geq \ldots \geq u_n$, with an overhead of $O(n \log n)$ time. Using the threshold function T_m^u defined in Eq. (2) and the job groups B_1, B_2^u, S^u formed by Eq. (3), the U-BBS algorithm for the $P \mid t_j = 1, 0 \leq p_j \leq u_j \mid C_{\max}$ problem can be presented as the following Algorithm 2.

Algorithm 2. U-BBS Algorithm

Input: B_1, B_2^u, S^u;
if $B_2^u = \emptyset$ **then**
  for $J_j \in B_1$ in increasing order of j **do**
    if $\frac{u_j}{t_j} < \varphi$ **then**
      schedule J_j untested on machine M_j;
    else
      test and schedule J_j on machine M_j;
    end if
  end for
else
  for $J_j \in B_1 \cup B_2^u$ in increasing order of j **do**
    test and schedule J_j on the least loaded machine;
  end for
end if
for $J_j \in S^u$ in increasing order of j **do**
  schedule J_j untested on the least loaded machine;
end for

Lemma 6. *If $B_2^u = \emptyset$, then for the job J_j, $j = 1, 2, \ldots, n$, we have*

$$\alpha_j \leq \begin{cases} \varphi, & J_j \in B_1, \\ T_m^u, & J_j \in S^u; \end{cases}$$

otherwise, for the job J_j, $j = 1, 2, \ldots, n$, we have

$$\alpha_j \leq \begin{cases} 1 + \frac{1}{T_m^u}, & J_j \in B_1 \cup B_2^u, \\ T_m^u, & J_j \in S^u. \end{cases}$$

Lemma 7. *If $J_n \in B_2^u$ and $m \geq 2$, then we have $C^B \leq \frac{(4m-1)T_m^u}{3m} C^*$.*

Proof. In this case, $|B_2^u| \neq \emptyset$. Because J_n is scheduled on the least-loaded machine, by Lemma 6 and Eq. (4), we have

$$\begin{aligned} C^B &\leq \frac{1}{m} \sum_{J_k \in \mathcal{J} \setminus J_n} p_k^B + p_n^B \leq \frac{1}{m} \sum_{J_k \in \mathcal{J}} p_k^B + \left(1 - \frac{1}{m}\right) p_n^B \\ &\leq \left(1 + \frac{1}{T_m^u}\right) \frac{1}{m} \sum_{J_k \in \mathcal{J}} \rho_k + \left(1 - \frac{1}{m}\right) p_n^B \\ &\leq \left(1 + \frac{1}{T_m^u}\right) C^* + \left(1 - \frac{1}{m}\right) p_n^B. \end{aligned}$$

If $p_n^B = p_n + 1 \leq C^*$, then by Lemma 2(ii),

$$C^B \leq \left(2 + \frac{1}{T_m} - \frac{1}{m}\right) C^* \leq \frac{(4m-1)T_m^u}{3m} C^*.$$

Therefore we can assume $p_n + 1 > C^*$. It means $\rho_n = u_n \leq C^*$ and $p_n^B \leq (1 + \frac{1}{T_m^u}) C^*$. Note that $u_n + 1 = p_n + 1 > C^*$ and $\rho_j \geq \tau_j \geq \tau_n = 1$. Hence in the optimal schedule, J_n must be assigned on a machine without other jobs. That is, $\mathcal{J} \setminus J_n$ will be scheduled on $m - 1$ machines. We distinguish two cases based on the value of m.

Case 1: $m \geq 3$. So we have $\frac{(4m-1)T_m^u}{3m} = 1 + \frac{1}{T_m^u} + \frac{4}{3}(1 - \frac{1}{m})$. So it is sufficient to demonstrate the upper bound is $1 + \frac{1}{T_m^u} + \frac{4}{3}(1 - \frac{1}{m})$.

Subcase 1.1: $C^* \geq 3$. Then we have

$$C^B \leq \left(1 + \frac{1}{T_m^u}\right) C^* + \left(1 - \frac{1}{m}\right)(u_n + 1) \leq \left(1 + \frac{1}{T_m^u} + \frac{4}{3}\left(1 - \frac{1}{m}\right)\right) C^*.$$

Subcase 1.2: $C^* < 3$. It is obvious that $|B_1 \cup B_2^u| \leq 2m - 1$. Otherwise, $|B_1 \cup B_2^u| \geq 2m$ and $|\mathcal{J} \setminus J_n| \geq 2m - 1$. Note that $\mathcal{J} \setminus J_n$ must be scheduled on $m - 1$ machines. So there exists a machine with at least 3 jobs on it, a contradiction.

(i) $m \geq 4$. If there exists a machine whose load is no more than $(\frac{4}{3} - \frac{4}{3m}) C^*$, then we have

$$C^B \leq \left(\frac{4}{3} - \frac{4}{3m}\right) C^* + \left(1 + \frac{1}{T_m^u}\right) C^* = \left(1 + \frac{1}{T_m^u} + \frac{4}{3}\left(1 - \frac{1}{m}\right)\right) C^*.$$

So we can assume the loads of all machines are at least $(\frac{4}{3}-\frac{4}{3m})C^*$ before J_n comes. By Lemma 2(iii), we have $\rho_{i_j} \geq \frac{\frac{4}{3}-\frac{4}{3m}}{1+\frac{1}{T_m^u}}C^* \geq \frac{2}{3}C^*$, $j=1,2,\ldots,\ell$. So in the optimal schedule, $J_{i_1}, J_{i_2}, \ldots, J_{i_\ell}, J_n$ must be scheduled on $\ell+1$ machines, respectively. Note that there are $m-\ell-1$ machines and at least $2(m-\ell)$ jobs left. So there is at least one machine with 3 jobs, a contradiction.

(ii) $m=3$. If there is a machine whose load is no more than $\frac{8}{9}C^*$ before J_n arrives, then by Lemma 6, we have

$$C^B \leq \frac{8}{9}C^* + \left(1+\frac{1}{T_m^u}\right)C^* = \left(\frac{17}{9}+\frac{1}{T_m}\right)C^*.$$

So we can assume that the loads of all machines exceed $\frac{8}{9}C^*$ before J_n comes. Since $B_2^u \leq 2$, then $\ell \geq 2$.

If $\ell=3$, then $|B_2^u|=1$. By Lemmas 2(i) and 6, we have $\rho_i \geq \frac{8}{9+\frac{9}{T_3}}C^* \geq \frac{1}{2}C^*$, $i=1,2,3$. So we have at least 3 jobs whose processing times are bigger than $\frac{1}{2}C^*$, a contradiction.

If $\ell=2$, then $|B_2^u|=2$. Let J_s be the job with less processing time in $\mathcal{J}^u$. If $u_s \leq \frac{2}{3}C^*$, then $u_n \leq u_s \leq \frac{2}{3}C^*$ and by Lemmas 2(iv) and 6, we have

$$C^B \leq \frac{2}{3}\left(2+\frac{2}{T_m^u}\right)C^* \leq \left(\frac{17}{9}+\frac{1}{T_m^u}\right)C^*.$$

So we can assume that $u_s > \frac{2}{3}C^*$. If $\rho_s = p_s+1$, then since $p_s^B > \frac{8}{9}C^*$, we have $\rho_s = p_s^B > \frac{8}{9}C^*$. Otherwise $\rho_s = u_s > \frac{2}{3}C^*$. Hence in the optimal solution, all the jobs of $\mathcal{J}^u$ must be assigned separately on two machines. Note that $|B_2^u|=2$. So there exists a job of B_2^u that is assigned on a machine with one job of $\mathcal{J}^u$, a contradiction.

Case 2: $m=2$. Recall that in the optimal solution, $\mathcal{J}\setminus J_n$ must be scheduled on $m-1$ machines. So we have $C^* \geq \sum_{i=1}^{n-1}\rho_i$ and by Lemma 6, $\sum_{i=1}^{n-1}p_i^B \leq (1+\frac{1}{T_2^u})C^*$. Since J_n is scheduled on the least-loaded machine, we obtain

$$C^B \leq \left(\frac{1}{2}+\frac{1}{2T_2^u}\right)C^* + \left(1+\frac{1}{T_2^u}\right)C^* = \frac{7T_2^u}{6}C^*.$$

□

Lemma 8. *If $J_n \in S^u$ and $m \geq 2$, then we have $C^B \leq \frac{(4m-1)T_m^u}{3m}C^*$.*

Theorem 2. *When $m \geq 2$, the competitive ratio of the U-BBS algorithm is at most $\frac{(4m-1)T_m^u}{3m}$, which is increasing in m and approaches $\frac{7+\sqrt{97}}{6} \approx 2.8081$.*

Proof. Case 1: $J_n \in B_1$. By Lemma 6 and Eq. (4), we have $C^B \leq \varphi\rho_n \leq \varphi C^*$.

Case 2: $J_n \in B_2^u$. By Lemma 7, we have $C^B \leq \frac{(4m-1)T_m^u}{3m}C^*$.

Case 3: $J_n \in S^u$. By Lemma 8, then we have $C^B \leq \frac{(4m-1)T_m^u}{3m}C^*$. □

5 Conclusion

In this paper, we investigate the non-preemptive multiprocessor scheduling with testing to minimize the makespan. We proposed online approximation algorithms BBS for the general testing case and U-BBS for the uniform testing case, respectively. Our algorithms adopt an earlier idea to test those jobs for which the testing operation is very cheap compared to the given processing time upper bound. The new ingredients are to sort the jobs in the non-increasing order of their τ-values, to partition the jobs differently into three groups, and to process the jobs group by group using the sorted job order. We showed that both our BBS and U-BBS algorithms are more competitive than the previous best algorithms, and their competitive ratios approach to 2.9513 and 2.8081, respectively, when the number of machines tends to infinity.

Albers and Eckl [2] proved a lower bound of $2 - \frac{1}{m}$ (and $\varphi \approx 1.6180$ the golden ratio, which is relevant only if $m = 1, 2$) on the competitive ratios, even in the uniform testing case. One sees the big gaps and thus further improved approximation algorithms are desired, as well as better lower bounds. In particular, when $m = 2$, both online and semi-online (with known maximum job processing time) multiprocessor scheduling problems admit optimal approximation algorithms, while there is a big gap of 0.695 and 0.653, respectively, for scheduling with testing in the general testing case and uniform testing case.

Acknowledgments. This research is supported by the NSERC Canada.

References

1. Albers, S., Eckl, A.: Explorable uncertainty in scheduling with non-uniform testing times. In: Kaklamanis, C., Levin, A. (eds.) WAOA 2020. LNCS, vol. 12806, pp. 127–142. Springer, Cham (2021). https://doi.org/10.1007/978-3-030-80879-2_9
2. Albers, S., Eckl, A.: Scheduling with testing on multiple identical parallel machines. In: Lubiw, A., Salavatipour, M. (eds.) WADS 2021. LNCS, vol. 12808, pp. 29–42. Springer, Cham (2021). https://doi.org/10.1007/978-3-030-83508-8_3
3. Cai, S.Y.: Semi online scheduling on three identical machines. J. Wenzhou Teach. Coll. 23, 1–3. (2002). (In Chinese)
4. Chen, B., Vliet, A., Woeginger, G.: New lower and upper bounds for on-line scheduling. Oper. Res. Lett. **16**, 221–230 (1994)
5. Dürr, C., Erlebach, T., Megow, N., Meißner, J.: Scheduling with explorable uncertainty. In: ITCS 2018, pp. 30:1–14. LIPIcs 94 (2018), https://doi.org/10.4230/LIPIcs.ITCS.2018.30
6. Faigle, U., Kern, W., Turan, G.: On the performance of on-line algorithms for partition problems. Acta Cybern. **9**, 107–119 (1989)
7. Fleischer, R., Wahl, M.: On-line scheduling revisited. J. Sched. **3**, 343–353 (2000)
8. Graham, R.L.: Bounds for certain multiprocessing anomalies. Bell Labs Tech. J. **45**, 1563–1581 (1966)
9. He, Y., Zhang, G.: Semi online scheduling on two identical machines. Computing **62**, 179–187 (1999)
10. Lee, K., Lim, K.: Semi-online scheduling problems on a small number of machines. J. Sched. **16**, 461–477 (2003)

11. Rudin III, J.F.: Improved bounds for the on-line scheduling problem. Ph.D. thesis (2001)
12. Rudin, J.F., III., Chandrasekaran, R.: Improved bound for the online scheduling problem. SIAM J. Comput. **32**, 717–735 (2003)
13. Tan, Z., Li, R.Q.: Pseudo lower bounds for online parallel machine scheduling. Oper. Res. Lett. **43**, 489–494 (2015)
14. Wu, Y., Huang, Y., Yang, Q.F.: Semi-online multiprocessor scheduling with the longest given processing time. J. Zhejiang Univ. Sci. Edn. **35**, 23–26 (2008). (in Chinese)

Author Index

Printed in the United States
by Baker & Taylor Publisher Services